AutoCAD 2021

3D Tutorials

Tutorial Books

For resource files, contact us at
Online.books999@gmail.com

Table of Contents

Introduction

Welcome to *AutoCAD 2012 3D Tutorials* book. This book is written to assist students, designers, and engineering professionals in designing 3D models. It covers the essential features and functionalities of AutoCAD using relevant tutorials and exercises.

Chapter 1: Getting Started with AutoCAD 3D

Introduction to AutoCAD

AutoCAD is a legendary software used to create 2D drawings and 3D Models. It enables you to convert 2D drawings into a 3D model rapidly. In addition to that, you can manipulate existing geometry easily. It makes it quick and simple for non-expert or casual CAD users to create new models and modify existing models.

System requirements

The following are system requirements for running AutoCAD smoothly on your system.

- Microsoft Windows 8.1, Windows 7 SP1, Windows 10 (64 - bit only).
- CPU Type:
    - Basic: 2.5 to 2.9 gigahertz (GHz)
    - Recommended: 3 gigahertz (GHz) or faster
- 8 GB of RAM (16 GB Recommended).
- Resolution 1920 x 1080 or higher recommended with True Color.
- Resolutions up to 3840 x 2160 supported on Windows 10, 64 bit systems (with capable display card) for High Resolution & 4K Displays.
- 7 GB of free space for installation.
- Google Chrome Browser.
- .NET Framework Version 4.8 or later

Starting AutoCAD 2021

To start **AutoCAD 2021**, double-click the **AutoCAD 2021** icon on your Desktop (or) click **Start > All apps > AutoCAD 2021 > AutoCAD 2021**.

3D Modeling Workspaces in AutoCAD

In AutoCAD, there are separate workspaces created to work on 3D models. In these workspaces, the tools are organized into ribbon tabs, menus, toolbars, and palettes to perform a specific task in 3D Modeling. You can activate these workspaces by using the **Workspace** drop-down located on the **Quick Access Toolbar**, or by using the **Workspace Switching** menu on the status bar. You can also start an AutoCAD session directly in the 3D Modeling workspace using the **acad3D.dwt, acadiso3D.dwt, acad -Named Plot Styles3D**, or **acadISO-Named Plot Styles3D** templates.

Tip: If the Workspace drop-down is not displayed at the top left corner, click the down arrow next to the Quick Access Toolbar. Next, select Workspace from the drop-down; the Workspace drop-down will be visible on the Quick Access Toolbar.

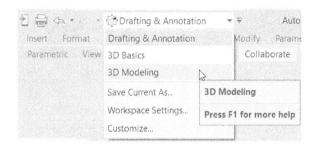

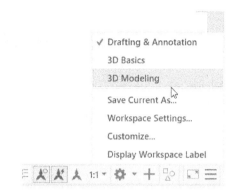

There are two workspaces of 3D Modeling: **3D Basics** and **3D Modeling**. The **3D Basics** workspace has commonly used tools, whereas the **3D Modeling** workspace includes all the tools required for creating 3D models.

The 3D Modeling Workspace

Activating the **3D Modeling** workspace either using the template or from the **Workspace** drop-down displays the screen as shown below. It contains the ribbon and tools related to 3D Modeling. By default, the **Home** tab is activated in the ribbon. From this tab, you can access the tools for creating and editing solids and meshes, modifying the model display, working with coordinate systems, sectioning 3D models, etc.

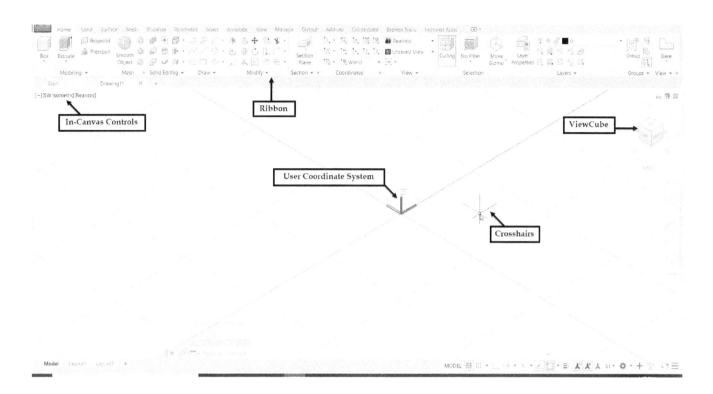

There are some additional tabs, such as **Solid**, **Surface**, **Mesh**, and **Render**. The **Solid** tab contains tools to create solid models; the **Surface** and **Mesh** tabs are used to create surface models and complex shapes; the **Visualize** tab is used for creating realistic images of solid and surface models.

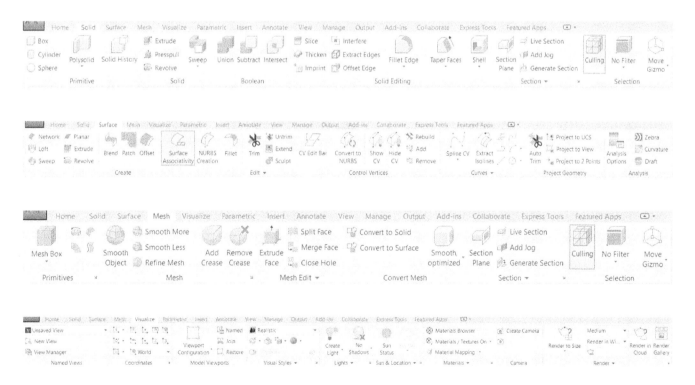

The **ViewCube** can be used to modify the view of the model quickly and easily. It is located at the top right corner of the graphics window. Using the ViewCube, you can switch between the standard and isometric views, rotate the model, switch to the **Home** view of the model, and create a new user coordinate system. You can also change the way the ViewCube functions by using the **ViewCube Settings** dialog. Right-click on the ViewCube, and then select the **ViewCube Settings** option; the **ViewCube Settings** dialog will be opened.

You can also modify the model view by using the In-canvas controls. In addition to that, you can also change the Visual Style of the model and control the display of other tools in the graphics window using the In-canvas controls.

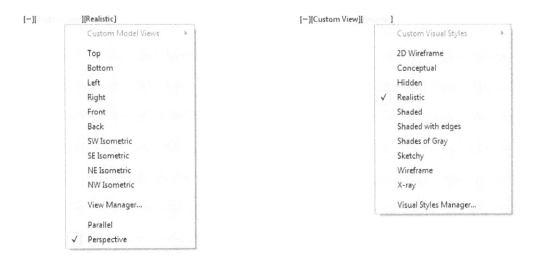

AutoCAD Help

Press F1 or type a keyword in the search bar located at the top right corner of the window to get help. On the **Autodesk AutoCAD 2021 –Help** window, click the **Find** option next to the topic; an animated arrow appears on the window showing the tool location.

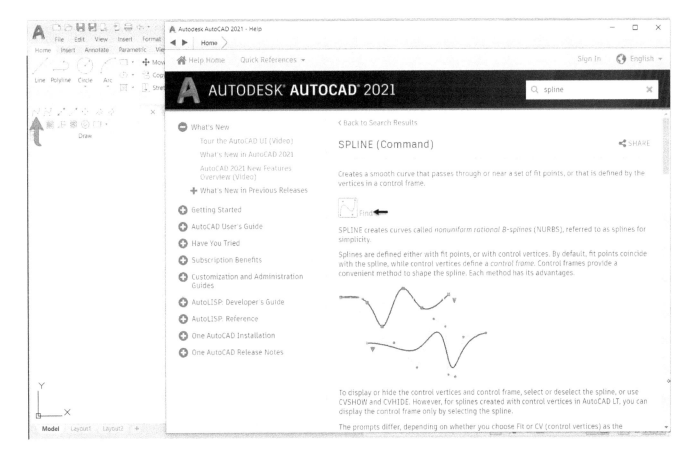

Chapter 2: Extrude and Revolve Features

This chapter covers the methods and commands to create extruded and revolved features.

Tutorial 1 (Millimeters)

In this example, you create the part shown below.

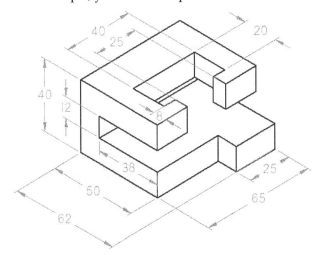

Creating a New File

1. Click the **AutoCAD 2021** icon on your desktop.
2. On the **Start** page, click **Get Started** drop-down > **acadiso3D.dwt**.

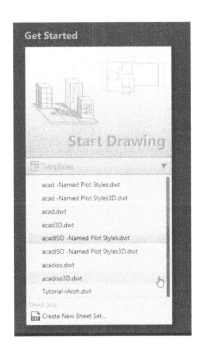

3. On the status bar, click **Workspace Switching** drop-down > **3D Modeling**.
4. Deactivate the **GRIDMODE** icon on the status bar.

Creating a Box

1. On the ribbon, click **Home** tab > **Modeling** panel > **Primitive** drop-down > **Box**.

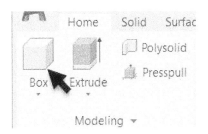

2. Type 0,0 in the command line and press ENTER. The first corner of the box is defined.
3. Type 50,65 in the command line and press ENTER. The second corner of the box is defined.
4. Type 40 in the command line and press ENTER. The height of the box is specified.
5. Click the **Zoom Extents** icon on the Navigation Bar to fit the box into the graphics window.

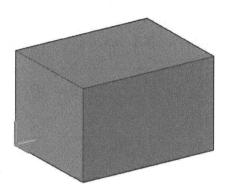

Creating the Cuts using the Subtract Boolean Operation

1. On the ribbon, click **Home** tab > **Modeling** panel > **Primitive** drop-down > **Box**.
2. Type 12,0,14 in the command line to specify the first corner of the box.
3. Press ENTER.
4. Select the **Length** option from the command line.
5. Click the ORTHOMODE icon on the status bar.
6. Move the pointer along the X direction, as shown.

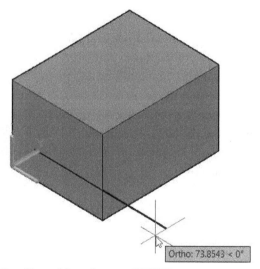

7. Type 38 and press ENTER.
8. Move the pointer along the Y direction, as shown.

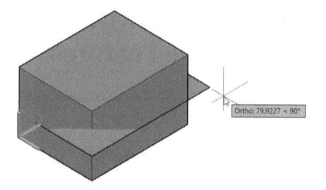

9. Type 65 and press ENTER.
10. Move the pointer upward.

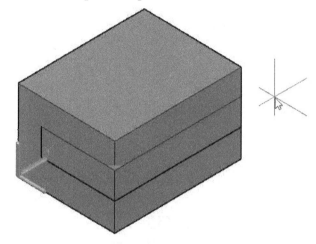

11. Type 12 and press ENTER.
12. On the ribbon, click **Home** tab > **View** panel > **Visual Style** drop-down > **Shades of Gray**.

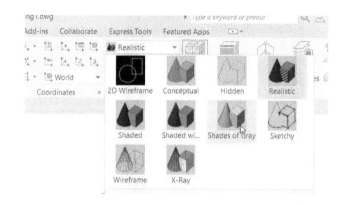

13. On the ribbon, click **Home** tab > **Solid Editing** panel > **Solid, Subtract**.

14. Select the large box and press ENTER.

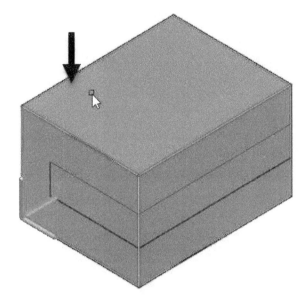

15. Select the small box and press ENTER.

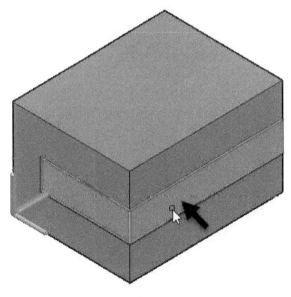

16. On the ribbon, click **Home** tab > **Modeling** panel > **Primitive** drop-down > **Box**.
17. Type 42,20,26 in the command line to specify the first corner of the box.
18. Select the **Length** option from the command line.

19. Click the ORTHOMODE icon on the status bar.
20. Move the pointer along the X direction, as shown.

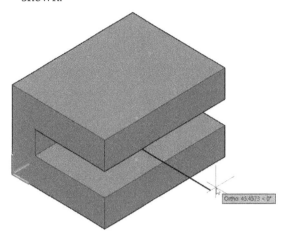

21. Type 8 and press ENTER.
22. Move the pointer along the Y direction, as shown.

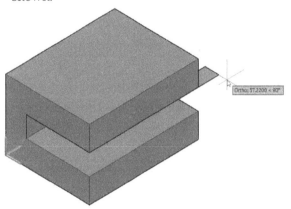

23. Type 25 and press ENTER.
24. Move the pointer upward.
25. Type 14 and press ENTER.

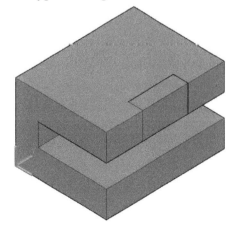

26. On the ribbon, click **Home** tab > **Modeling** panel > **Primitive** drop-down > **Box**.
27. Type 22,12.5,26 in the command line to specify the first corner of the box.
28. Select the **Length** option from the command line.
29. Click the ORTHOMODE icon on the status bar.
30. Move the pointer along the X direction, as shown.

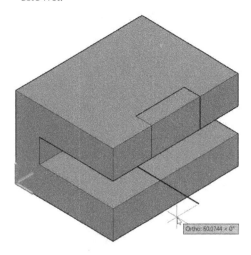

31. Type 20 and press ENTER.
32. Move the pointer along the Y direction, as shown.

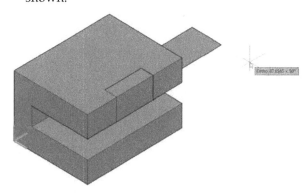

33. Type 40 and press ENTER.
34. Move the pointer upward.
35. Type 14 and press ENTER.

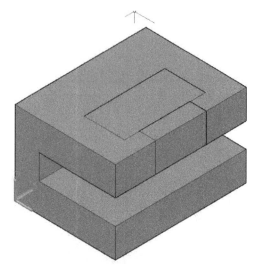

36. On the ribbon, click **Home** tab > **Solid Editing** panel > **Solid, Subtract**.
37. Select the main body and press ENTER.

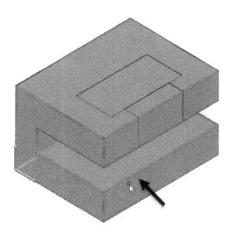

38. Select the two boxes and press ENTER.

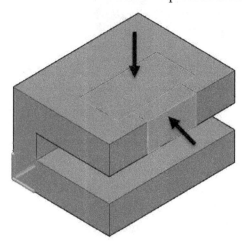

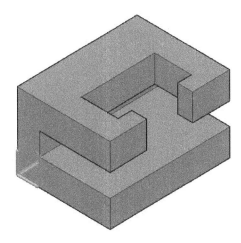

Creating a Solid using the Union Boolean Operation

1. On the ribbon, click **Home** tab > **Modeling** panel > **Primitive** drop-down > **Box**.
2. Type 50,40 in the command line and press ENTER. The first corner of the box is defined.
3. Type 62,65 in the command line and press ENTER. The second corner of the box is defined.
4. Type 14 in the command line and press ENTER. The height of the box is specified.

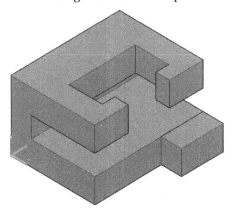

5. On the ribbon, click **Home** tab > **Solid Editing** panel > **Solid, Union**.

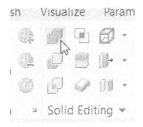

6. Select the main body.

7. Select the small box and press ENTER.

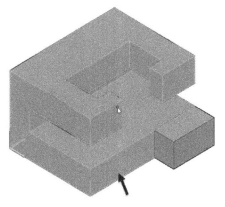

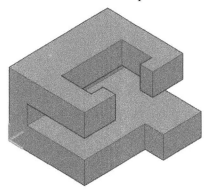

8. Click the **Save** icon on the Quick Access Toolbar.

9. Browse to a location on your computer.
10. Type **Ch2_tutorial1** in the **File name** box.
11. Click the **Save** button.
12. Close the file tab located above the graphics window.

Tutorial 2 (Inches)

In this example, you create the part shown below.

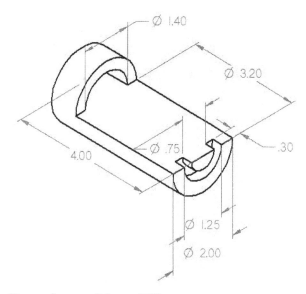

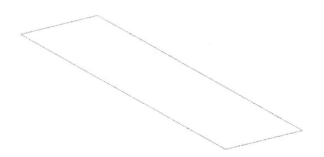

Creating a New File

1. Click the **AutoCAD 2021** icon on your desktop.
2. On the **Start** page, click **Get Started** drop-down > **acad3D.dwt**.
3. On the status bar, click **Workspace Switching** drop-down > **3D Modeling**.
4. Deactivate the **GRIDMODE** icon on the status bar.

Creating a Revolved Solid

1. On the ribbon, click **Home > Draw > Rectangle**.

2. Type 0,0 in the command line and press ENTER. The first corner of the rectangle is defined.
3. Select **Dimensions** from the command line.
4. Type **4** in the command line and press ENTER. The length of the rectangle is defined.
5. Type **1** in the command line and press ENTER. The width of the rectangle is defined.
6. Move the pointer toward the left and click to create the rectangle.
7. Click the **Zoom Extents** icon on the Navigation Bar to fit the rectangle in the graphics window.

8. On the ribbon, click **Home > Modeling > Solids drop-down > Revolve**.

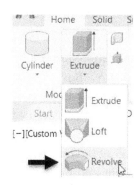

9. Select the rectangle and press ENTER.
10. Select the **X** option from the command line. The X-axis is used as the axis of revolution.
11. Type 180 in the command line and press ENTER. The angle of revolution is defined.

Creating a Cut feature

1. Activate the Dynamic UCS icon on the status bar.

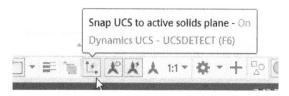

2. On the ribbon, click **Home > Draw > Circle** drop-down **> Circle, Diameter**.

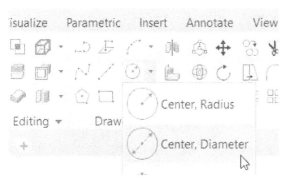

3. Place the pointer on the right face of the model; the face is highlighted.

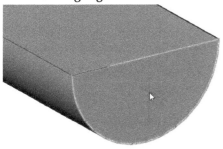

4. Click on the right face of the model.
5. Select the **Diameter** option from the command line.
6. Type 0.75 in the command line and press ENTER.

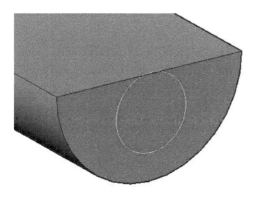

7. On the status bar, click the down arrow next to the **Object Snap** icon and select the **Center** and **Midpoint** options.
8. On the ribbon, click **Home > Draw > Circle** drop-down **> Circle, Diameter**.
9. Place the pointer on the right face of the model.

10. Click near the center point of the circle; the centerpoint of the circle is selected.

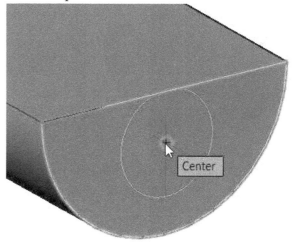

11. Type 1.25 in the command line and press ENTER.
12. Click on the two newly created circles; the Move gizmo is displayed at the centerpoint of the circles.
13. Click on the origin point of the Move gizmo.
14. Move the pointer upward and select the Midpoint of the horizontal edge.

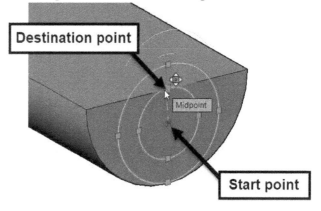

The circles are moved to the new location.

15. On the ribbon, click **Home > Modeling > Presspull**.

16. Click in the region between the two circles.

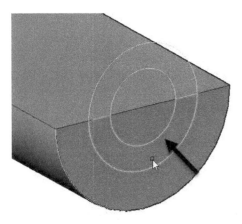

17. Type **-0.3** in the command line and press ENTER.

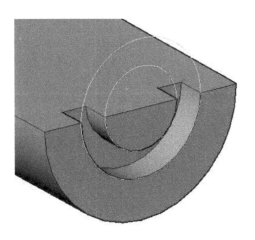

18. On the ribbon, click **Home > Draw > Rectangle**.
19. Select the left corner point of the model, as shown.

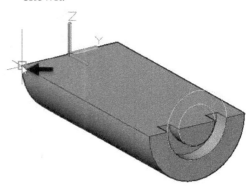

20. Select the **Dimensions** option from the command line.
21. Type 0.8 in the command line and press ENTER.
22. Type 0.3 in the command line and press ENTER.
23. Move the pointer toward the right and click to create the rectangle.

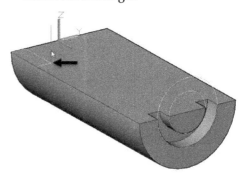

24. On the ribbon, click **Home > Modeling > Solid** drop-down > **Revolve**.
25. Select the rectangle and press ENTER.
26. Select the **X** option from the command line. The X-axis is selected as the axis of revolution.
27. Select the **Reverse** option from the command line. The revolution direction is reversed.
28. Type 180 in the command line and press ENTER.

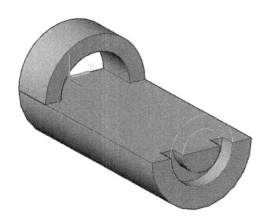

29. On the ribbon, click **Home** tab > **Solid Editing** panel > **Solid, Union**.
30. Select the main body.
31. Select the newly created revolved solid and press ENTER.

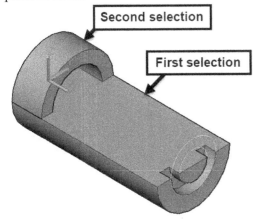

32. Select the two circles.

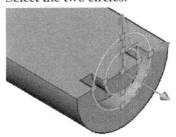

33. Right-click and select **Isolate > Hide Objects**.

34. Click **Application Menu > Save**.
35. Browse to a location on your computer.
36. Type **Ch2_tutorial2** in the **File name** box.
37. Click the **Save** button.
38. Click **Application Menu > Close > Current Drawing**.

Exercises

Exercise 1

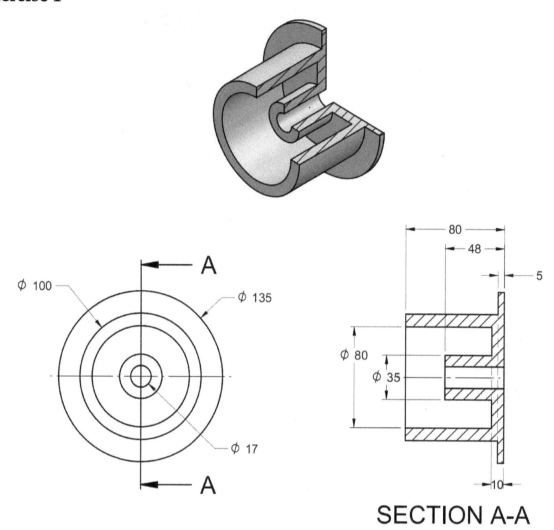

A

Φ 100

Φ 135

Φ 17

A

80

48

5

Φ 80

Φ 35

10

SECTION A-A

Exercise 2

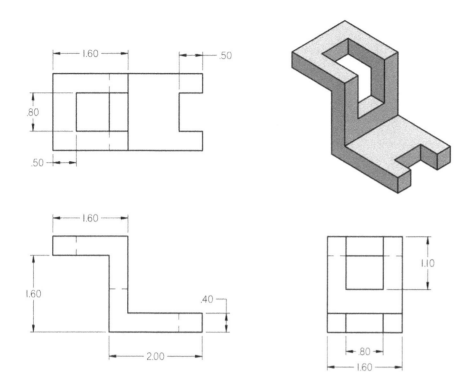

Exercise 3

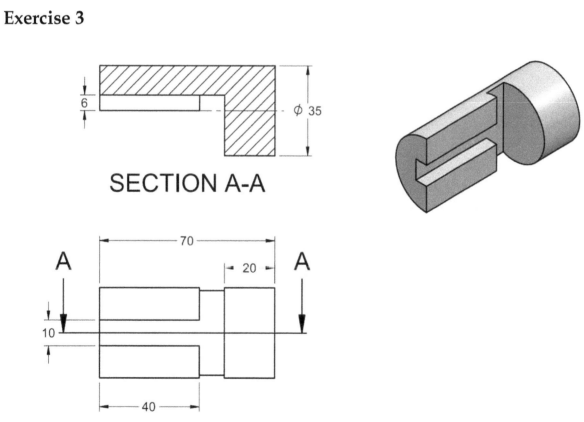

SECTION A-A

Chapter 3: Placed Features

Tutorial 1 (Millimetres)

In this example, you create the part shown below.

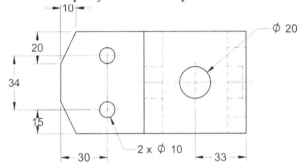

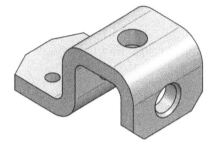

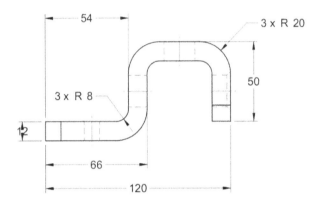

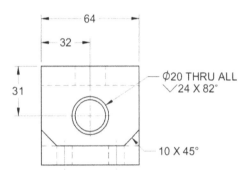

Creating a New File

1. Click the **AutoCAD 2021** icon on your desktop.
2. On the **Start** page, click **Get Started** drop-down > **acadiso3D.dwt**.
3. Deactivate the **GRIDMODE** icon on the status bar.

Creating the Extruded Solid

1. On the ribbon, click **Home** tab > **View** panel > **View Manager** drop-down > **Front**. The view orientation is changed to Front. Also, the X and Y axis of the UCS is set to Front view.

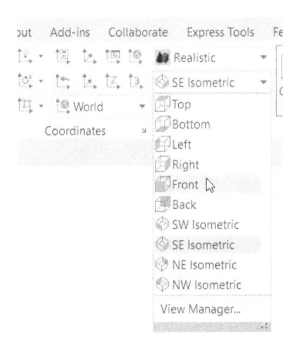

2. On the ribbon, click **Home** tab > **Draw** panel > **Polyline**.

Draw ▾

3. Type 0,0 in the command line and press ENTER.
4. Turn ON the Dynamic Input and ORTHOMODE icons on the status bar.
5. Move the pointer toward the right.
6. Type 66 and press ENTER.
7. Move the pointer upward.
8. Type 55 and press ENTER.
9. Move the pointer toward the right.
10. Type 42 and press ENTER.
11. Move the pointer downward.
12. Type 38 and press ENTER.
13. Press ESC.

14. On the ribbon, click the **Home** tab > **Modify** panel > **Offset**.
15. Type 12 in the command line and press ENTER. The offset distance is defined.
16. Select the polyline from the graphics window.
17. Move the pointer upward and click.

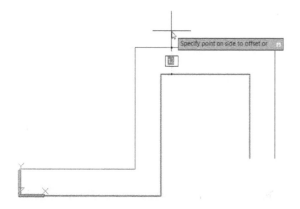

18. On the ribbon, click **Home** tab > **Draw** panel > **Line**.

Draw ▾

19. Select the right endpoints of the two polylines, as shown.

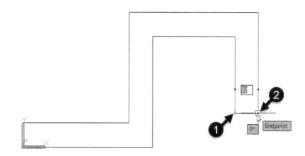

20. Press ESC.
21. On the ribbon, click **Home** tab > **Draw** panel > **Line**.
22. Select the left endpoints of the two polylines, as shown.

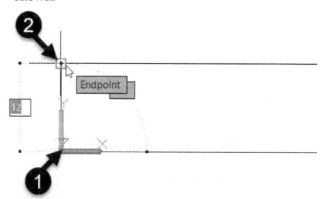

23. Press ESC.
24. On the **Home** tab of the ribbon, expand the **Draw** panel and click the **Region** icon.

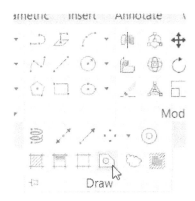

25. Create a selection window from left to right across all the elements of the drawing.

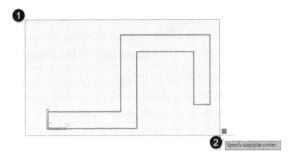

26. Press ENTER to convert all the 2D elements into a region.

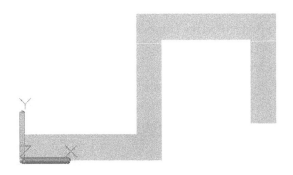

27. On the ribbon, click **Home** tab > **View** panel > **View Manager** drop-down > **SE Isometric**. The view orientation is changed to South East Isometric.

28. On the ribbon, click **Home** tab > **Modeling** panel > **Solids** drop-down > **Extrude**.

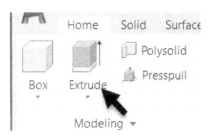

29. Click on the newly created region.
30. Press ENTER.
31. Move the pointer toward the right.
32. Type 64 and press ENTER.

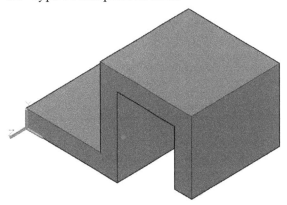

Creating Holes

1. On the ribbon, click **Home** tab > **Coordinates** panel > **UCS**.

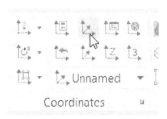

2. Select the lower-left corner of the right-side face to define the origin of the UCS.

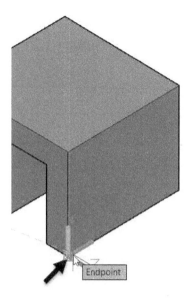

3. Move the pointer toward the right and select the lower right corner.

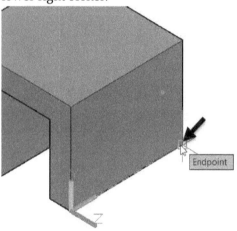

4. Move the pointer upward and select the top-left corner.

5. On the ribbon, click **Home** tab > **Modeling** panel > **Primitives** drop-down > **Cylinder**.

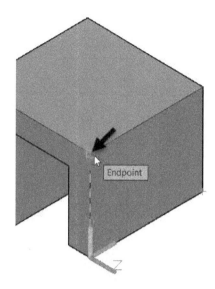

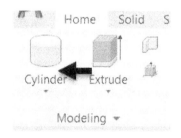

6. Type **32,19** in the command line and press ENTER. The centerpoint of the cylinder is defined.
7. Type **10** in the command line and press ENTER. The radius of the cylinder is defined.
8. Move the pointer toward the left and click to create the cylinder.

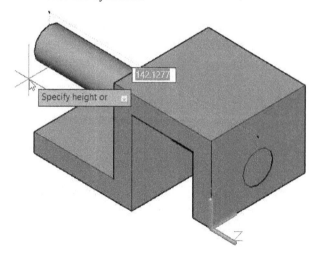

9. On the ribbon, click **Home** tab > **Solid Editing** panel > **Solid, Subtract** .
10. Select the main body and press ENTER.

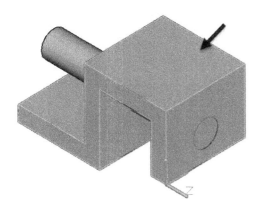

11. Select the cylinder and press ENTER. The cylinder is removed from the main body.

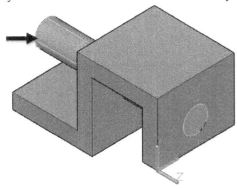

12. On the ribbon, click the **Solid** tab > **Solid Editing** panel > **Fillet Edge** drop-down > **Chamfer Edge**.

13. Select the circular edge of the hole.

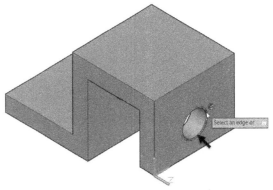

14. Select the **Distance** option from the command line.
15. Type **2** in the command line and press ENTER. Distance 1 is defined.
16. Type **1.5** in the command line and press ENTER. Distance 2 is defined.
17. Press ENTER twice to create the chamfer.

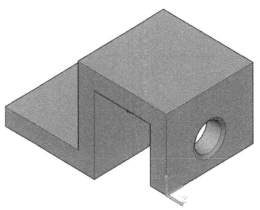

18. On the ribbon, click **Home** tab > **Coordinates** panel > **UCS** .
19. Select the top left corner of the right-side face to define the origin of the UCS.

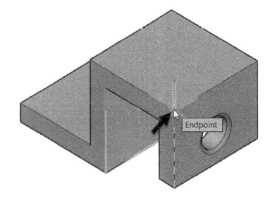

20. Move the pointer toward the right and select the top right corner of the right-side face.

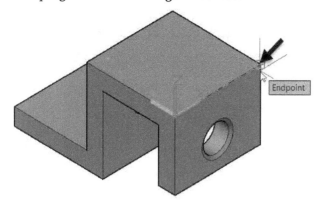

21. Move the pointer toward the left and select the corner of the top face, as shown.

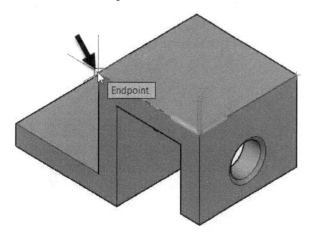

22. On the ribbon, click **Home** tab > **Modeling** panel > **Primitives** drop-down > **Cylinder**.
23. Type 32, 33 in the command line and press ENTER. The centerpoint of the cylinder is defined.
24. Type 10 in the command line and press ENTER to specify the radius of the cylinder.
25. Move the pointer downward and click to create the cylinder.

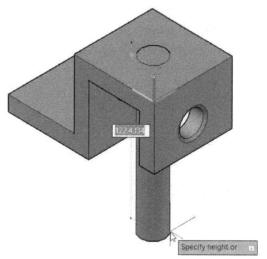

26. On the ribbon, click **Home** tab > **Coordinates** panel > **UCS** .
27. Select the front left corner of the flat face, as shown.

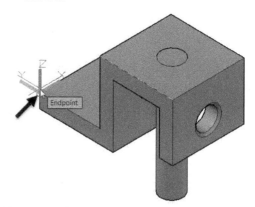

28. Move the pointer backward and select the corner point, as shown.

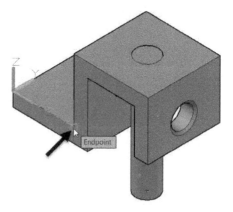

29. Move the pointer toward the right and select the midpoint of the edge, as shown.

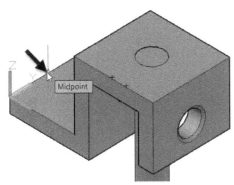

30. On the ribbon, click **Home** tab > **Modeling** panel > **Primitives** drop-down > **Cylinder**.

31. Type 30, 15 in the command line and press ENTER. The centerpoint of the cylinder is defined.

32. Type 5 in the command line and press ENTER to specify the radius of the cylinder.

33. Move the pointer downward and click to create the cylinder.

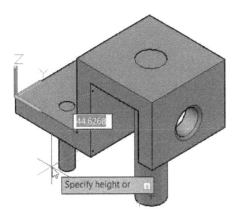

34. On the **Home** tab of the ribbon, expand the **Modify** panel and click the **Mirror** icon.

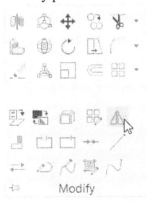

35. Select the newly created cylinder and press ENTER.

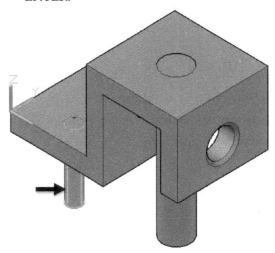

36. Select the midpoint of the horizontal edge of the top face, as shown.

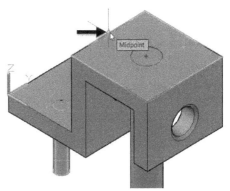

37. Move the pointer backward and select the midpoint of the horizontal edge.

38. Select **No** to keep the source object.

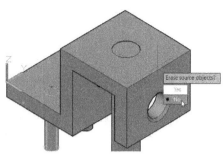

39. Click the front left corner of the ViewCube to change the orientation.

40. On the ribbon, click **Home** tab > **Solid Editing** panel > **Solid, Subtract**.
41. Select the main body and press ENTER.

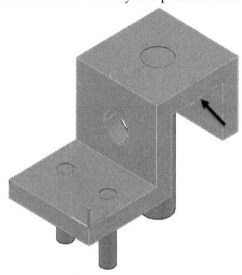

42. Select the three cylinders and press ENTER. The cylinders are removed from the main body.

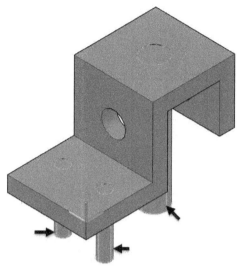

Creating Chamfers and Fillets

1. On the ribbon, click **Solid** tab > **Solid Editing** panel > **Fillet Edge** drop-down > **Chamfer Edge**.
2. Click on the vertical edge of the model, as shown.

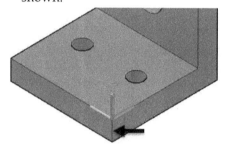

3. Select the **Distance** option from the command line.
4. Type **10** in the command line and press ENTER. Distance 1 is defined.
5. Type **20** in the command line and press ENTER. Distance 2 is defined.
6. Press ENTER twice to create the chamfer.

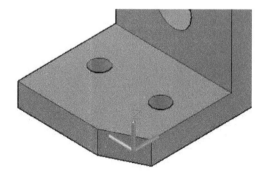

7. On the ribbon, click **Solid** tab > **Solid Editing** panel > **Fillet Edge** drop-down > **Chamfer Edge** .

8. Select the left vertical edge of the model, as shown.

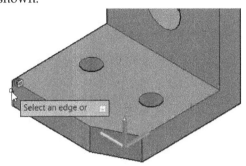

9. Select the **Distance** option from the command line.
10. Type **20** in the command line and press ENTER. Distance 1 is defined.
11. Type **10** in the command line and press ENTER. Distance 2 is defined.
12. Press ENTER twice to create a chamfer.

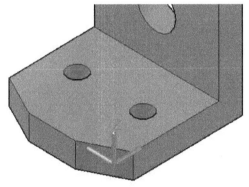

13. On the ribbon, click **Solid** tab > **Solid Editing** panel > **Fillet Edge**.

14. Click on the horizontal edge of the geometry, as shown.

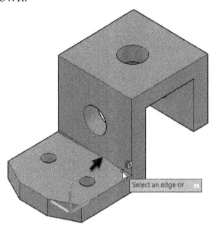

15. Click the bottom front corner of the ViewCube; the view orientation is changed.

16. Select the inner horizontal edge of the model, as shown.

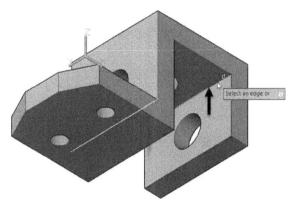

17. Select the bottom-back corner of the ViewCube. The orientation of the model is changed.

18. Select the visible inner horizontal edge.

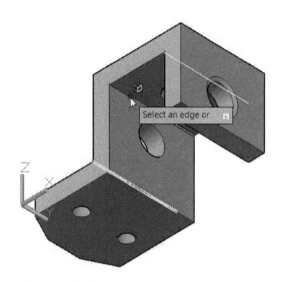

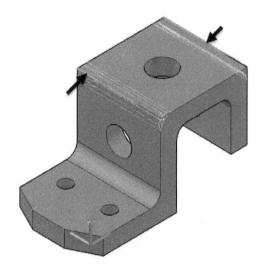

19. Select the **Radius** option from the command line.
20. Type 8 and press ENTER thrice.
21. Change the view orientation to SW Isometric.

24. Click on the bottom right corner of the ViewCube.

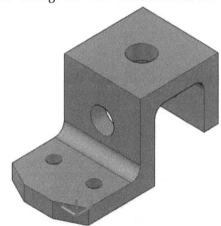

25. Select the bottom horizontal edge of the model.

22. On the ribbon, click **Solid** tab > **Solid Editing** panel > **Fillet Edge**.
23. Select the horizontal edges of the model, as shown.

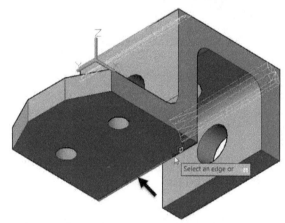

26. Select the **Radius** option from the command line.
27. Type 20 and press ENTER thrice.
28. Change the view orientation to SE Isometric.

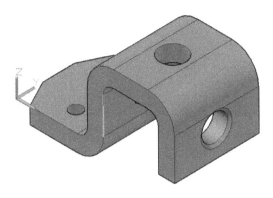

29. Click the **Orbit** tool on the Navigation Bar located at the right of the graphics window.

30. Press and hold the left mouse button and drag the pointer upward; the bottom portion is displayed.

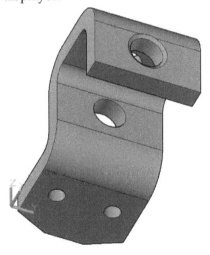

31. Right-click and select **Exit** from the shortcut menu.
32. On the ribbon, click **Solid** tab > **Solid Editing** panel > **Fillet Edge** drop-down > **Chamfer Edge** .
33. Click on the horizontal edges of the model, as shown.

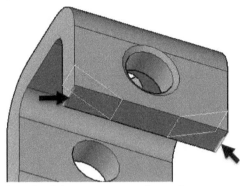

34. Select the **Distance** option from the command line.
35. Type 10 and press ENTER to specify the Distance 1.
36. Press ENTER to specify the Distance 2.
37. Press ENTER twice to create the chamfers.
38. Change the view orientation to SE Isometric.

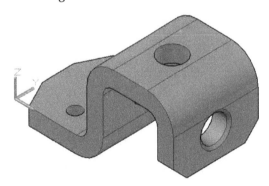

39. Click **Application Menu > Save**.
40. Browse to a location on your computer.
41. Type **Ch3_tutorial1** in the **File name** box.
42. Click the **Save** button.
43. Click **Application Menu > Close > Current Drawing**.

Exercises
Exercise 1 (Millimetres)

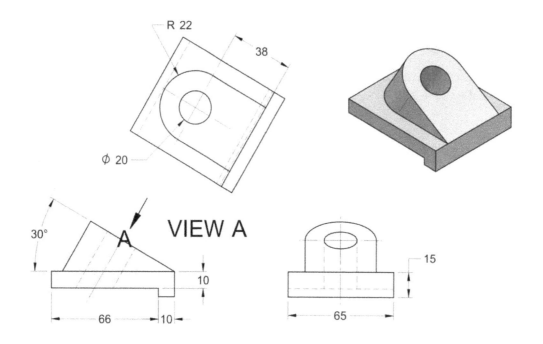

R 22

38

Ø 20

30°

A

VIEW A

10

66

10

15

65

Exercise 2 (Inches)

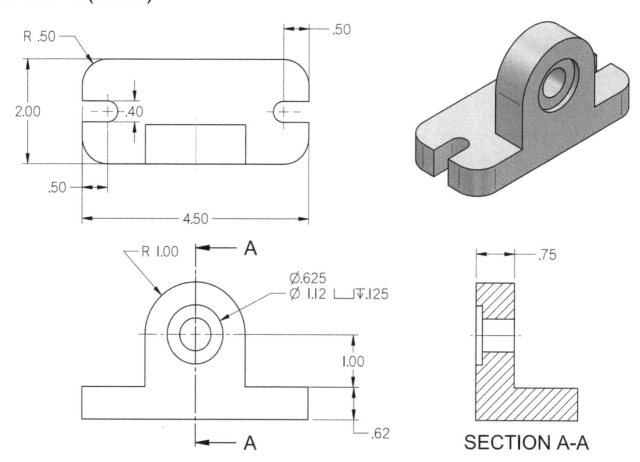

R .50

.50

2.00

.40

.50

4.50

R 1.00

A

Ø.625
Ø 1.12 ⌴ ⌵.125

1.00

.62

A

.75

SECTION A-A

Chapter 4: Patterned Geometry

Tutorial 1

In this example, you create the part shown next.

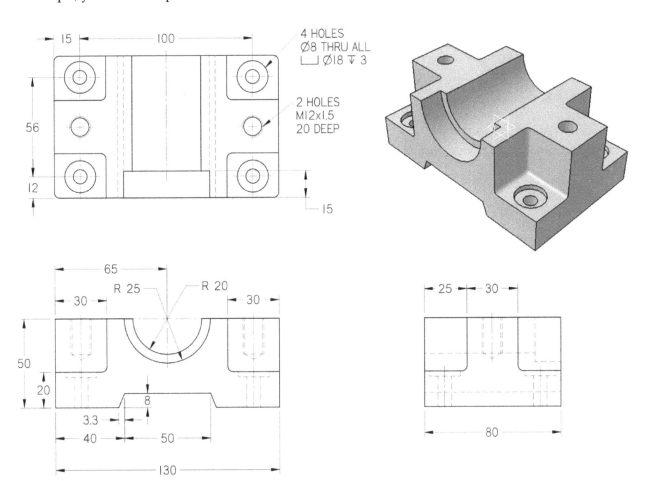

Creating a New File

1. Click the **AutoCAD 2021** icon on your desktop.
2. On the **Start** page, click **Get Started** drop-down > **acadiso3D.dwt**.
3. Deactivate the **GRIDMODE** icon on the status bar.

Creating the Box

1. On the ribbon, click **Home** tab > **Modeling** panel > **Primitive** drop-down > **Box**.
2. Type 0,0 in the command line and press ENTER. The first corner of the box is defined.

3. Make sure that the Dynamic Input icon is active on the status bar.
4. Type 130 in the length box and press the TAB key.
5. Type 80 and press ENTER.

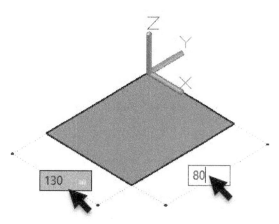

6. Move the pointer upward.
7. Type 50 and press ENTER to create the box.

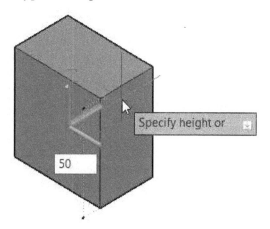

8. Deactivate the **Dynamic Input** icon on the Status bar.
9. Activate the **Orthomode** icon on the Status bar.
10. On the ribbon, click **Home** tab > **Modeling** panel > **Primitive** drop-down > **Box**.
11. Select the top left corner of the box, as shown.

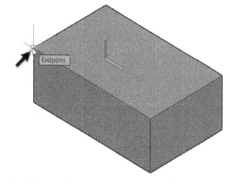

12. Select the **Length** option from the command line.
13. Type 30 and press ENTER.
14. Type -25 and press ENTER to define the width.

15. Type -30 and press ENTER to define the height.

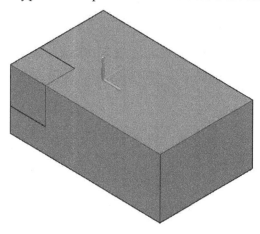

16. Change the **Visual Style** to **Shades of Gray**.

17. Place the pointer on any one of the edges of the small box.
18. Click when the small box is highlighted.

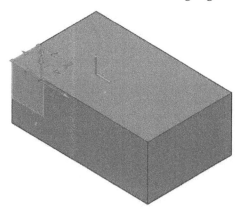

19. On the ribbon, **Home** tab > **Modify** panel > **Array** drop-down > **Rectangular Array**.

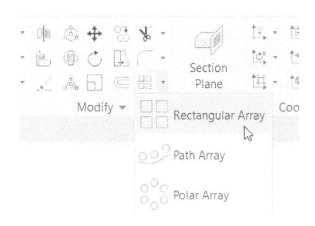

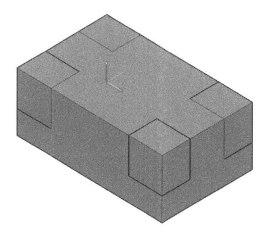

20. On the **Array Creation** tab, type **2** in the **Columns** and **Rows** boxes, respectively.
21. Type **100** in the **Between** box available on the **Columns** panel.

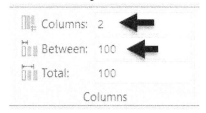

22. Type **55** in the **Total** box available on the **Rows** panel.

23. Deactivate the **Associative** icon on the ribbon.
24. Click the **Close Array** button on the ribbon.

25. On the ribbon, click **Home** tab > **View** panel > **Visual Style** drop-down > **2D Wireframe**.

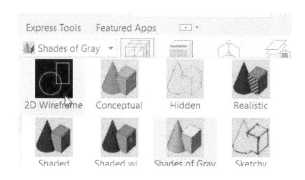

26. On the ribbon, click **Home** tab > **Solid Editing** panel > **Solid, Subtract**.
27. Click on the edge of the large box; the large box is selected.
28. Press ENTER.

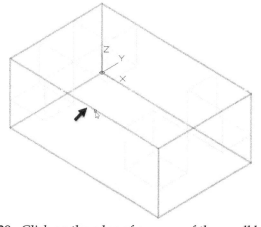

29. Click on the edge of any one of the small boxes.

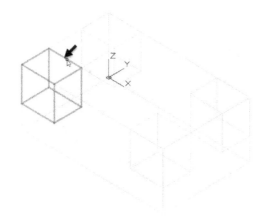

30. Likewise, select the remaining boxes and press ENTER.

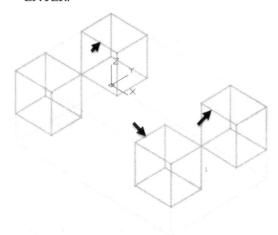

31. In the In-canvas controls, change the **Visual Style** to **Shades of Gray**.

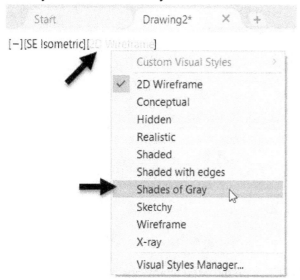

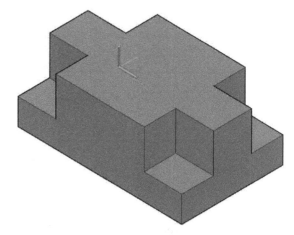

Creating the Hole features

1. On the ribbon, click **Home** tab > **Coordinates** panel > **UCS**.

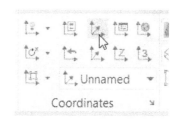

2. Select the lower-left corner of the subtraction to define the origin of the UCS.

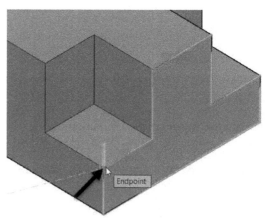

3. Move the pointer toward the right and select the corner point, as shown. The X-axis of the UCS is defined.

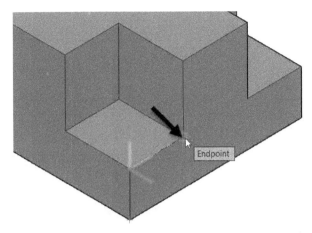

4. Move the pointer in the forward direction and click to define the Y-axis.

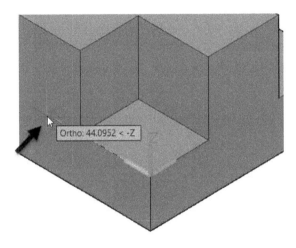

5. On the ribbon, click **Home** tab > **Modeling** panel > **Primitives** drop-down > **Cylinder**.
6. Type 12, 15 in the command line and press ENTER. The centerpoint of the cylinder is defined.
7. Type 4 in the command line and press ENTER.
8. Move the pointer downward and click to create the cylinder.

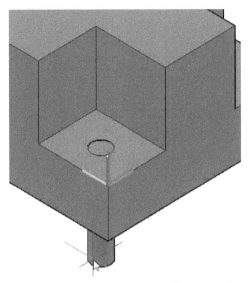

9. On the ribbon, click **Home** tab > **Modeling** panel > **Primitives** drop-down > **Cylinder**.
10. Type 12, 15 in the command line and press ENTER.
11. Type 9 in the command line and press ENTER.
12. Type -3 and press ENTER.

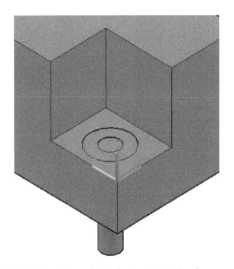

13. Set the **Visual Style** to **2D Wireframe**.

14. Select the two cylinders, as shown.

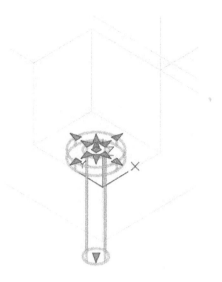

15. On the ribbon, **Home** tab > **Modify** panel > **Array** drop-down > **Rectangular Array**.
16. On the **Array Creation** tab, type **2** in the **Columns** and **Rows** boxes, respectively.
17. Type **56** in the **Between** box available on the **Columns** panel.

18. Type **100** in the **Between** box available on the **Rows** panel.

19. Deactivate the **Associative** icon on the ribbon.
20. Click the **Close Array** icon on the ribbon.

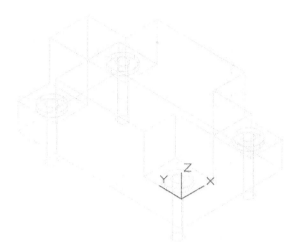

21. On the ribbon, click **Home** tab > **Solid Editing** panel > **Solid, Subtract**.
22. Click on the edge of the main body; the main body is selected.
23. Press ENTER.

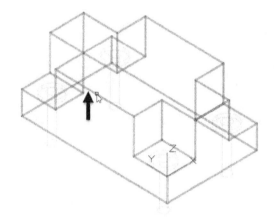

24. Select all the cylinders and press ENTER.

25. Change the **Visual Style** to **Shades of Gray**.

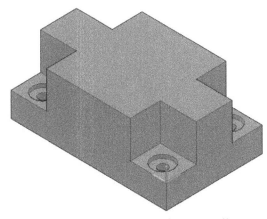

26. On the ribbon, click **Home** tab > **Coordinates** panel > **UCS**.

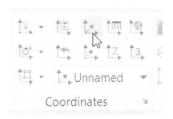

27. Select the corner point, as shown.

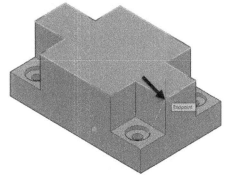

28. Move the pointer toward the right and click. The X-axis of the UCS is defined.

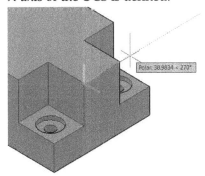

29. Move the pointer in the forward direction and click to define the Y-axis.

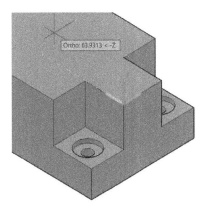

30. On the ribbon, click **Home** tab > **Modeling** panel > **Primitives** drop-down > **Cylinder**.
31. Type 15, 15 in the command line and press ENTER.
32. Type 6 in the command line and press ENTER.
33. Type -20 and press ENTER.

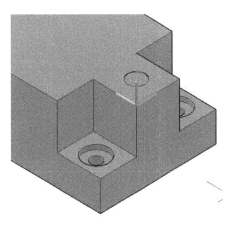

34. On the Status bar, click the down-arrow next to the **Object Snap** icon and select the **Midpoint** option.
35. Set the **Visual Style** to **2D Wireframe**.

36. Select the cylinder, as shown.

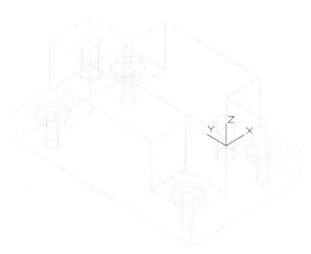

37. On the **Home** tab of the ribbon, expand the **Modify** panel and click the **Mirror** icon.
38. Select the midpoint of the front edge, as shown.

41. On the **Home** tab of the ribbon, click **Coordinates > Z-Axis Vector**.

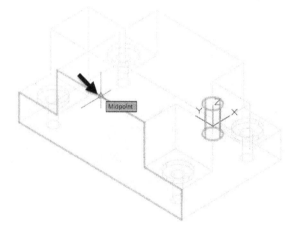

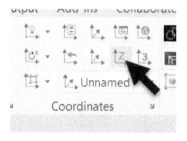

42. Select the corner point of the model, as shown.

39. Move the pointer downward and select the midpoint of the bottom edge.

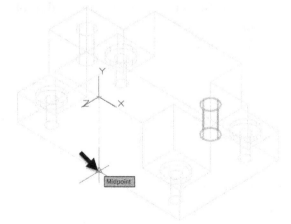

43. Move the pointer toward the right and select the corner point, as shown. The Z-axis of the UCS is defined.

40. Select the **No** option from the command line.

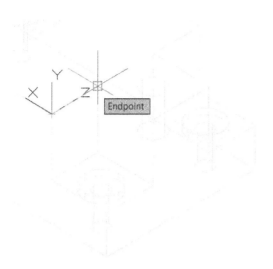

44. On the ribbon, click **Home** tab > **Modeling** panel > **Primitives** drop-down > **Cylinder**.
45. Select the midpoint of the front edge, as shown.

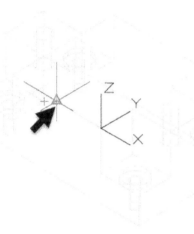

46. Type 20 in the command line and press ENTER.
47. Move the pointer toward the right and select the midpoint of the back edge, as shown.

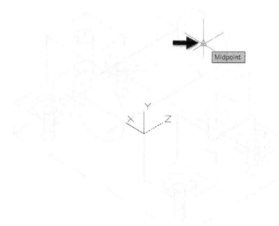

48. Deactivate the **Dynamic UCS** icon on the status bar.
49. On the ribbon, click **Home** tab > **Modeling** panel > **Primitives** drop-down > **Cylinder**.
50. Select the midpoint of the front edge, as shown.

51. Type 25 in the command line and press ENTER.
52. Move the pointer toward the right.
53. Type 15 and press ENTER.

54. On the ribbon, click **Home** tab > **Solid Editing** panel > **Solid, Subtract**
55. Click on the edge of the main body; the main body is selected.
56. Press ENTER.

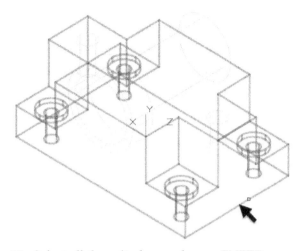

57. Select all the cylinders and press ENTER.

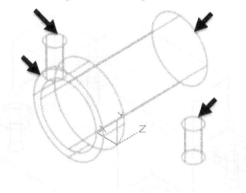

58. Set the **Visual Style** to **Shades of Gray**.

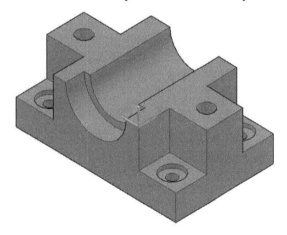

Creating the Presspull feature

1. On the ribbon, click **Home** tab > **Coordinates** panel > **UCS**.

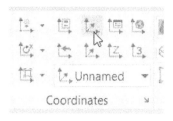

2. Select the lower-left corner of the model to define the origin of the UCS.

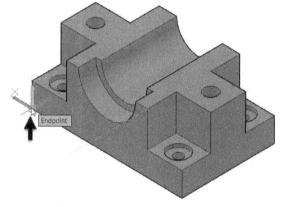

3. Move the pointer backward and select the lower right corner of the model.

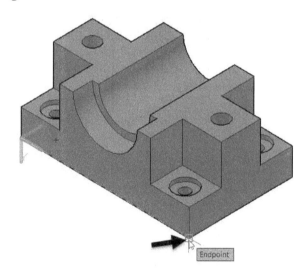

4. Move the pointer upwards and select the corner point, as shown.

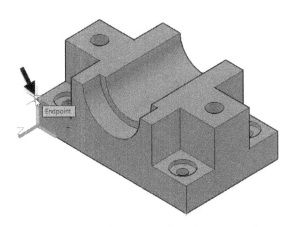

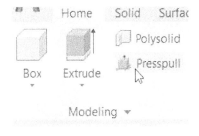

5. On the ribbon, click **Home** tab > **Draw** panel > **Polyline**.

6. Type **36.7, 0** in the command line, and press ENTER.
7. Type **40,8** in the command line and press ENTER.
8. Type **90,8** in the command line and press ENTER.
9. Type **93.3,0** in the command line and press ENTER.
10. Select the **Close** option from the command line.

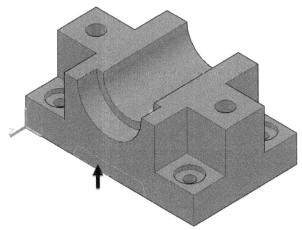

11. On the ribbon, click **Home** tab > **Modeling** panel > **Presspull**.

12. Click in the region bounded by the closed polyline.

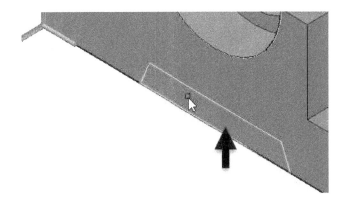

13. Type -80 and press ENTER.

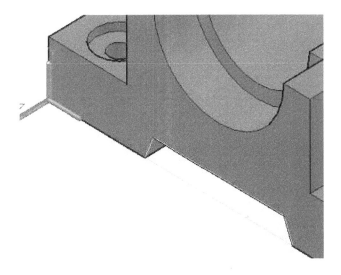

14. Press ESC.
15. Set the **Visual Style** to **2D Wireframe**.
16. On the ribbon, click **Solid** tab > **Solid Editing** panel > **Fillet Edge**.

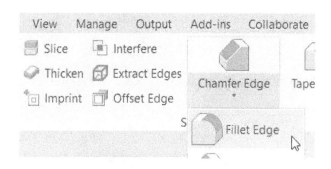

17. Select the inner edges of the subtractions, as shown.

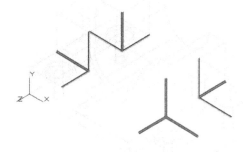

18. Press ENTER.

19. Select the **Radius** option from the command line.
20. Type **2** in the command line and press ENTER.
21. Press ENTER twice to create the fillets.
22. Set the **Visual Style** to **Shades of Gray**.

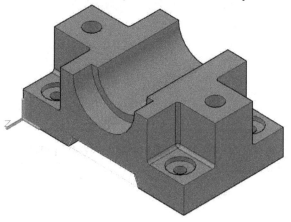

23. Click the **Save** icon on the Quick Access Toolbar.
24. Type **Ch4_tutorial1** in the **File Name** box.
25. Click the **Save** button.
26. Click **Application Menu > Close > Current Drawing**.

Exercises

Exercise 1

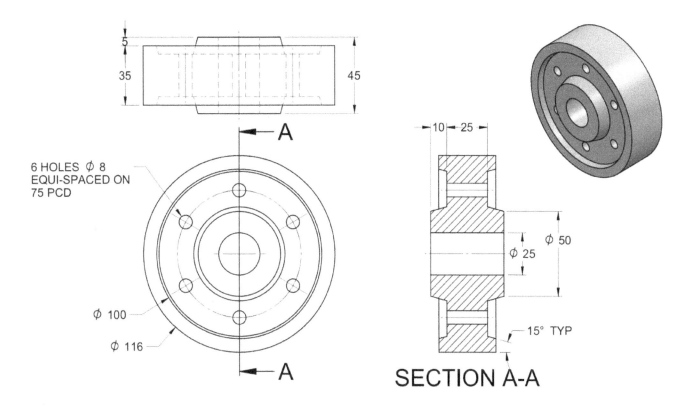

6 HOLES Ø 8
EQUI-SPACED ON
75 PCD

Ø 100

Ø 116

SECTION A-A

Chapter 5: Sweep Features

Tutorial 1

In this example, you create the part shown below.

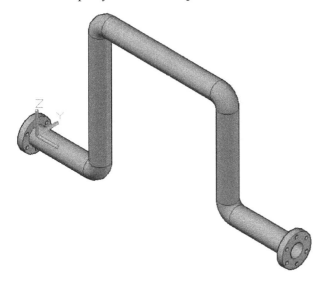

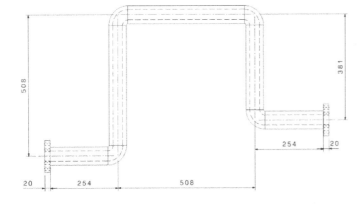

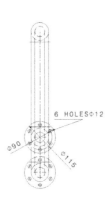

PIPE I.D. 51

PIPE O.D. 65

Creating a New File

1. Click the **AutoCAD 2021** icon on your desktop.
2. On the **Start** page, click **Get Started** drop-down > **acadiso3D.dwt**.
3. Deactivate the **GRIDMODE** icon on the status bar.

Creating the Swept Solid

1. On the ribbon, click **Home** tab > **View** panel > **Restore View** drop-down > **Front**.

2. On the ribbon, click **Home** tab > **Draw** panel > **Polyline**.
3. Type 0,0 in the command line and press ENTER. The first point of the polyline is defined.
4. Make sure that the **Dynamic Input** and **Orthomode** icons are active on the status bar.
5. Move the pointer toward the right.
6. Type 254, and press ENTER.
7. Move the pointer upward.
8. Type 508, and press ENTER.
9. Move the pointer toward the right.
10. Type 508, and press ENTER.
11. Move the pointer downward.
12. Type 381, and press ENTER.
13. Move the pointer toward the right.
14. Type 254, and press ENTER.
15. Press ESC.

16. On the ribbon, click **Home** tab > **Modify** panel > **Fillet** drop-down > **Fillet**.

17. Select the **Radius** option from the command line.
18. Type **38** in the command line and press ENTER.
19. Select Vertical and horizontal lines meeting at the corner, as shown.

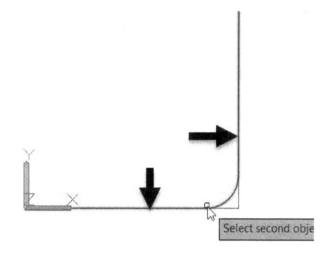

20. Press ENTER to activate the **Fillet** command.
21. Select the vertical and horizontal lines meeting at the corner, as shown.

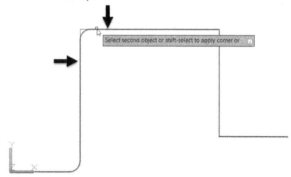

22. Likewise, create fillets at the corners, as shown.

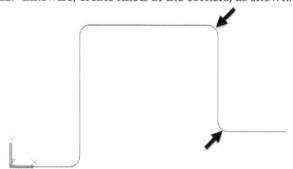

23. Change the view orientation to SE Isometric.

24. On the ribbon, click **Home** tab > **Coordinates** panel > **Z-Axis Vector**.

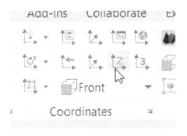

25. Select the endpoint of the polyline.

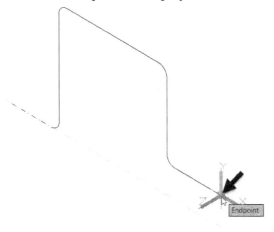

26. Move the pointer along the horizontal line and select its endpoint, as shown.

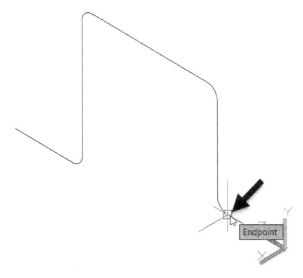

27. On the ribbon, click **Home** tab > **Draw** panel > **Circle** drop-down > **Circle, Diameter**.

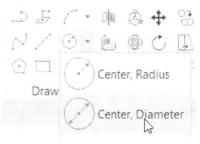

28. Type 0,0 and press ENTER.
29. Type 65 and press ENTER.

30. On the ribbon, click the **Home** tab > **Modeling** panel > **Solids** drop-down > **Sweep**.

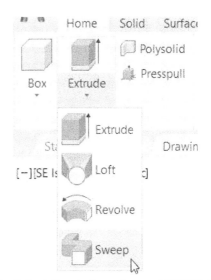

31. Select the circle and press ENTER to define the object to sweep.
32. Select the polyline to define the path.

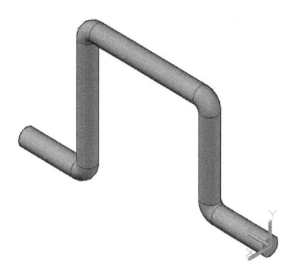

Shelling the Solid

1. On the ribbon, click **Solids** tab > **Solid Editing** panel > **Shell**.

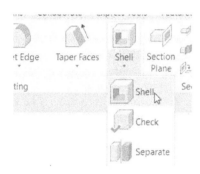

2. Click on the swept solid.
3. Click on the front end of the swept solid.

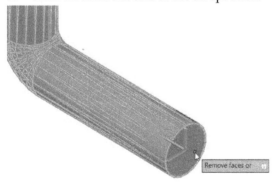

4. On the Navigation Bar, click the **Orbit** icon.
5. Press and hold the left mouse button and drag the pointer toward the right.
6. Scroll the mouse wheel backward.
7. Right-click and select Exit.

8. Click on the back end face of the swept solid.

9. Press ENTER to accept the selection.
10. Type 7 in the command line and press ENTER. The shell thickness is defined.
11. Select **eXit** from the command line.
12. Select **eXit** from the command line.
13. Change the view orientation to SE Isometric.

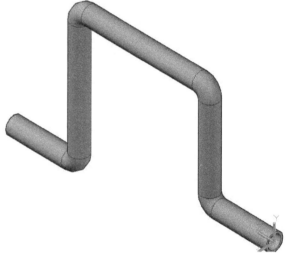

Adding the Flange

1. On the ribbon, click **Home** tab > **Draw** panel > **Circle** drop-down > **Circle, Diameter**.

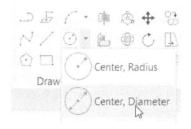

2. Type 0,0 and press ENTER.
3. Type 115, and press ENTER.

4. Activate the **Dynamic Input** icon on the status bar.
5. On the ribbon, click **Home** tab > **Modeling** panel > **Presspull**.
6. Click on the end face of the swept solid.

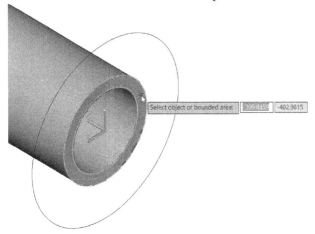

7. Move the pointer toward the right.
8. Type 20 and press ENTER.

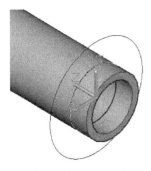

9. Click inside the circle and move pointer toward the right.
10. Type 20 and press ENTER.

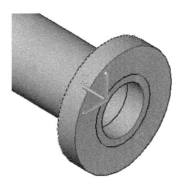

Creating the Circular Pattern

1. On the ribbon, click **Home** tab > **Modeling** panel > **Primitives** drop-down > **Cylinder**.
2. Type 0, 45 in the command line and press ENTER. The centerpoint of the cylinder is defined.
3. Type 6 in the command line and press ENTER.
4. Move the pointer toward the right.
5. Type 20 and press ENTER to create the cylinder.

6. Set the **Visual Style** to **2D Wireframe**.

7. Select the small cylinder, as shown.

8. On the ribbon, **Home** tab > **Modify** panel > **Array** drop-down > **Polar Array**.

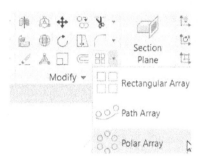

9. Select the centerpoint of the press pulled solid; the centerpoint of the polar array is defined.

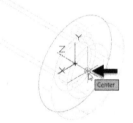

10. On the **Array Creation** tab, type **6** and **360** in the **Items** and **Fill** boxes, respectively.

11. Deactivate the **Associative** icon on the ribbon.
12. Click the **Close Array** icon on the ribbon.

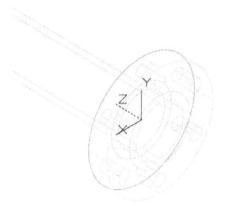

13. On the ribbon, click **Home** tab > **Solid Editing** panel > **Solid, Subtract**.
14. Click on the edge of the press pulled solid; the press pulled solid is selected.
15. Press ENTER.

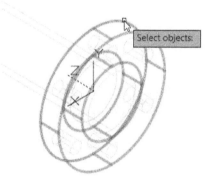

16. Select all the cylinders of the polar array and press ENTER.

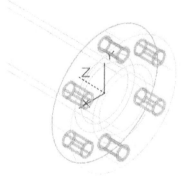

17. Change the **Visual Style** to **Shades of Gray**.

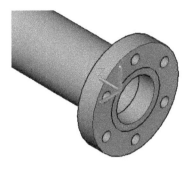

18. Select the two press pulled solids, as shown.

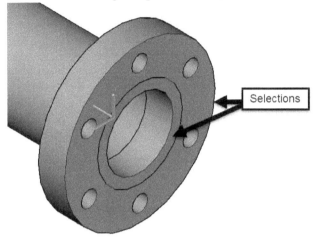

19. On the ribbon, **Home** tab > **Modify** panel > **Copy**.

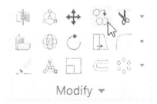

20. Select the centerpoint of the press pulled solid to define the base point.

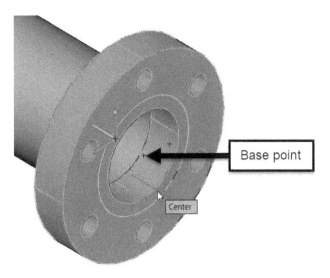

21. Move the pointer in the forward direction.
22. Select the centerpoint of the circular edge on the backside, as shown.

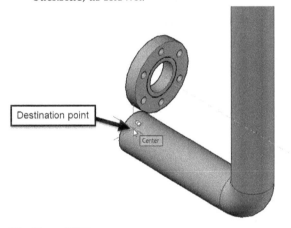

23. Press ESC.
24. On the ribbon, click **Home** tab > **Solid Editing** panel > **Solid, Union**.

25. Create a selection window across the entire model.

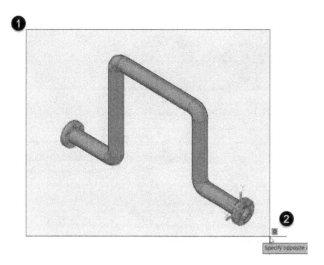

26. Press ENTER. All the solids are combined into a single solid.
27. Click the **Save** icon on the Quick Access Toolbar.
28. Type **Ch5_tutorial1** in the **File Name** box.
29. Click the **Save** button.
30. Click **Application Menu > Close > Current Drawing**.

Exercises
Exercise1

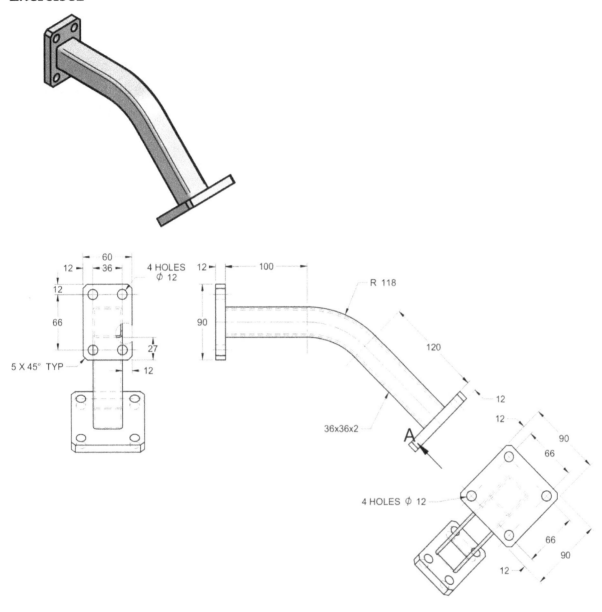

Chapter 6: Loft Features

Tutorial 1

In this example, you create the part shown below.

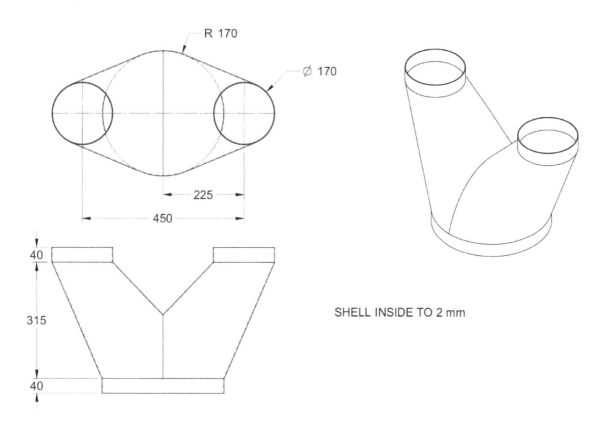

SHELL INSIDE TO 2 mm

Creating a New File

1. Click the New icon on the Quick Access Toolbar.

2. Select the **acadiso3D** template from the **Select template** dialog.
3. Click the **Open** button.
4. Deactivate the **GRIDMODE** icon on the status bar.

Creating a Loft Feature

1. On the ribbon, click **Home** tab > **Modeling** panel > **Primitives** drop-down > **Cylinder**.

2. Type 0, 0 in the command line and press ENTER. The centerpoint of the cylinder is defined.
3. Type 170 in the command line and press ENTER.
4. Move the pointer upward.
5. Type 40 and press ENTER to create the cylinder.

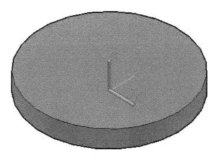

6. On the ribbon, click **Home** tab > **Coordinates** panel > **Origin**.
7. Type 0,0,315 in the command line and press ENTER. The UCS is moved up to 315 distance in the Z-direction.

12. Click on the top face of the cylinder.

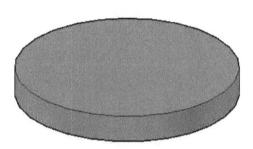

8. On the ribbon, click **Home** tab > **Draw** panel > **Circles** drop-down > **Circle, Diameter**.
9. Type 225,0 in the command line and press ENTER.
10. Type 170 in the command line and press ENTER. The diameter of the circle is specified.

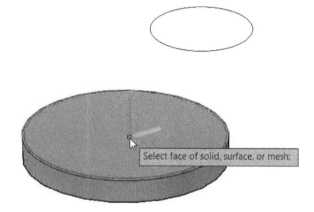

13. Press ENTER to accept the selection. The UCS is placed on the selected face.
14. On the ribbon, click **Home** tab > **Draw** panel > **Circles** drop-down > **Circle, Diameter**.
15. Place the pointe on the circular edge of the cylinder; the center point of the cylinder is highlighted.
16. Select the center point of the cylinder.

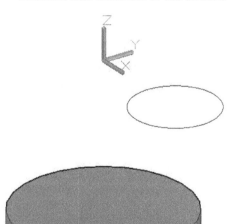

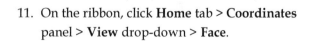

11. On the ribbon, click **Home** tab > **Coordinates** panel > **View** drop-down > **Face**.

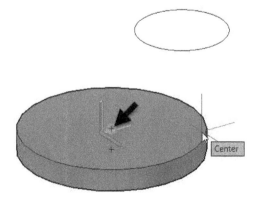

17. Type 340 in the command line and press ENTER.
18. On the ribbon, click the **Home** tab > **Modeling** panel > **Solids** drop-down > **Loft**.
19. Select the two circles.

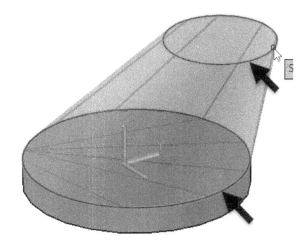

20. Press ENTER twice to create the *Loft* feature.

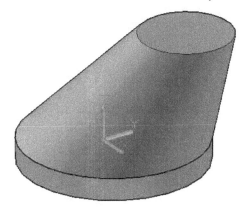

Extruding a Planar Face of the Model

1. On the ribbon, click the **Home** tab > **Solid Editing** panel > **Edit Faces** drop-down > **Extrude Faces**.

2. Click on the top face of the *Loft* solid.

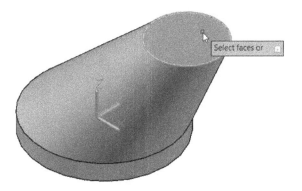

3. Press ENTER to accept the selection.
4. Type 40 and press ENTER to specify the extrusion height.
5. Press ENTER to accept 0 as the taper angle.
6. Press ENTER twice to extrude the face.

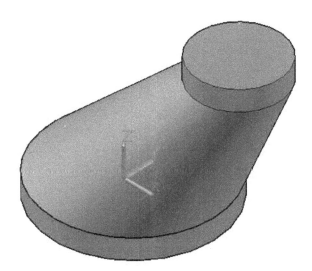

Mirroring the Loft solid.

1. On the status bar, click the down arrow next to the **Object Snap** icon, and then select the **Quadrant** option.

59

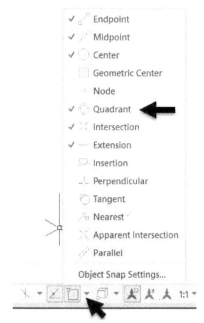

2. Select the loft solid.

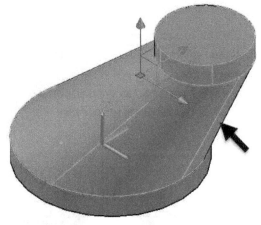

3. On the ribbon, click **Home** tab > **Modify** panel > **3D Mirror**.

4. Select the **YZ** option from the command line. The orientation of the mirror plane is defined

5. Select the quadrant point of the circular edge, as shown. The location of the mirror plane is defined.

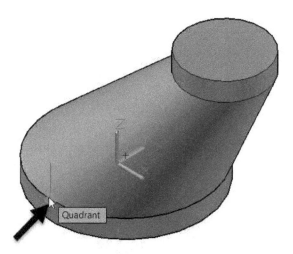

6. Select **No** from the command line to keep the original object.

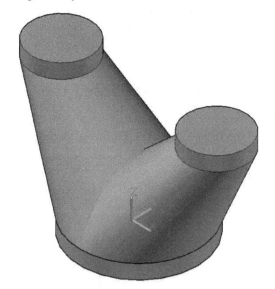

7. On the ribbon, click **Home** tab > **Solid Editing** panel > **Solid, Union**.

8. Create a selection window across all the elements of the model.

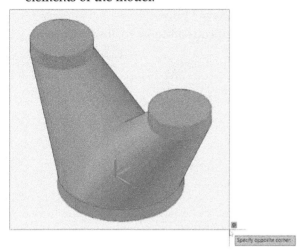

9. Press ENTER to combine all the objects.

Shelling the Model geometry

1. On the ribbon, click **Solids** tab > **Solid Editing** panel > **Shell**.

2. Click on the model.
3. Click on the top faces of the model.

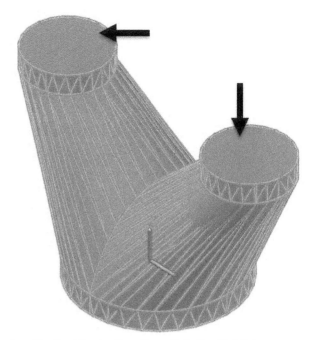

4. On the Navigation Bar, click the **Orbit** icon.
5. Press and hold the left mouse button and drag the pointer upward.
6. Right-click and select Exit.
7. Click on the bottom face of the model.

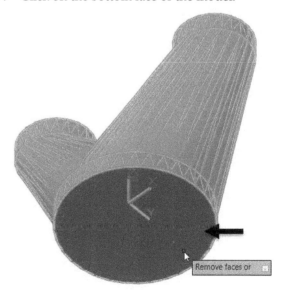

8. Press ENTER to accept the selection.
9. Type 2 in the command line and press ENTER. The shell thickness is defined.
10. Select **eXit** from the command line.
11. Select **eXit** from the command line.
12. Change the view orientation to SE Isometric.

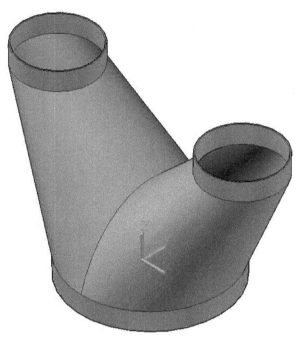

13. Click the **Save** icon on the Quick Access Toolbar.
14. Type **Ch6_tutorial1** in the **File Name** box.
15. Click the **Save** button.
16. Click **Application Menu > Close > Current Drawing**.

Tutorial 2 (Inches)

In this example, you will create the part shown below.

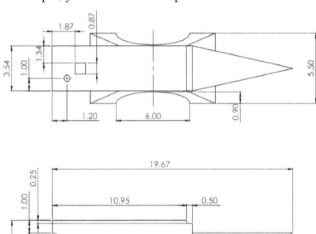

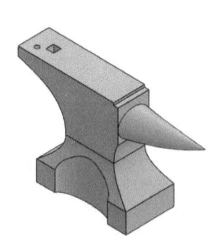

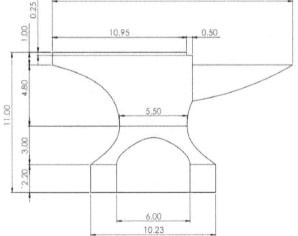

1. Start **Autodesk AutoCAD 2021**.

2. On the **Application Menu**, click the **New** icon; the **Select Template** dialog appears.

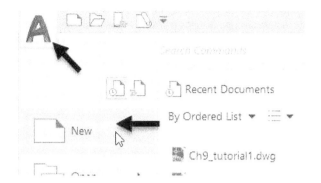

3. On this dialog, click **acad3D**, and then click the **Open** button.
4. Deactivate the **GRIDMODE** icon on the status bar.
5. On the ribbon, click **Home** tab > **Modeling** panel > **Primitive** drop-down > **Box**.

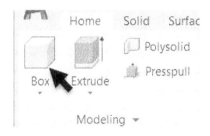

6. Type 0,0 in the command line and press ENTER. The first corner of the box is defined.
7. Type 10.23, 5.5 in the command line, and press ENTER. The second corner of the box is defined.
8. Type 2.2 in the command line and press ENTER. The height of the box is specified.
9. Click the **Zoom Extents** icon on the Navigation Bar to fit the box into the graphics window.

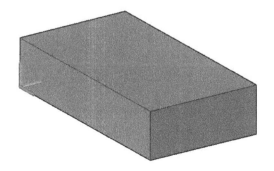

Create the Lofted solid

1. Change the **Visual Style** to **2D Wireframe**.
2. On the ribbon, click **Home** tab > **Selection** panel > **Filter** drop-down > **Edge**.

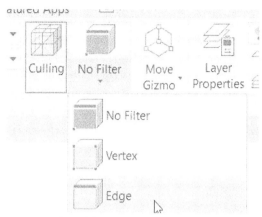

3. On the ribbon, click **Home** tab > **Solid Editing** panel > **Edge** drop-down > **Extract Edges**.

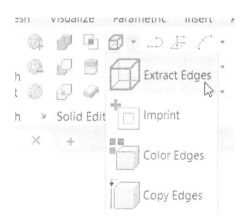

4. Select the edges of the model, as shown.

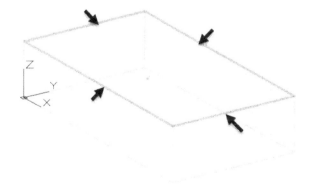

5. Press ENTER.
6. On the ribbon, click the **Home** tab > **Modify** panel > **Offset**.

7. Type 2.365, and press ENTER.
8. Select the extracted left edge.
9. Move the pointer toward the right and click.

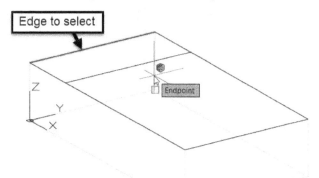

10. Select the extracted right edge.
11. Move the pointer toward the left and click.

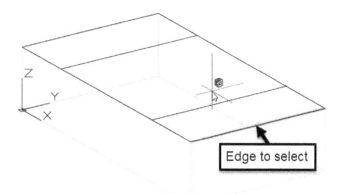

12. Press ENTER twice.
13. Type 0.98 and press ENTER.
14. Select the front edge.
15. Move the pointer backward and click.

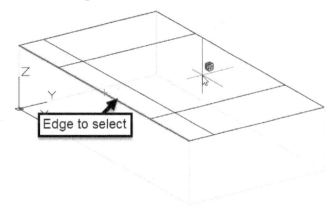

16. Select the back edge.
17. Move the pointer forward and click.

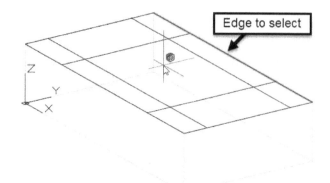

18. On the ribbon, click **Home** tab > **Modify** panel > **Trim/Extend** drop-down > **Trim**.
19. Select the portions of the offset lines, as shown.

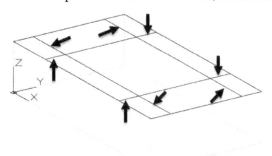

20. On the **Home** tab of the ribbon, expand the **Modify** panel and click the **Edit Polyline** icon.

21. Select the **Multiple** option from the command line.
22. Select the trimmed lines, as shown.

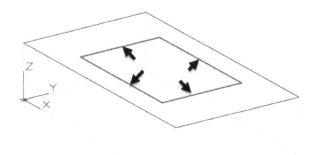

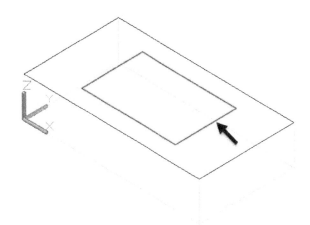

23. Press ENTER.
24. Select the **Yes** option from the command line.
25. Select the **Join** option from the command line.
26. Press ENTER.
27. On the **Home** tab of the ribbon, expand the **Modify** panel and click the **Edit Polyline** icon.
28. Select the **Multiple** option from the command line.
29. Select the extracted lines, as shown.

38. Select the Z-axis of the move gizmo.

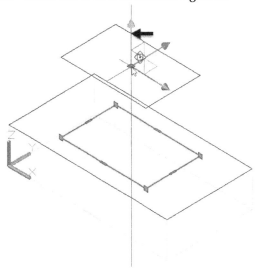

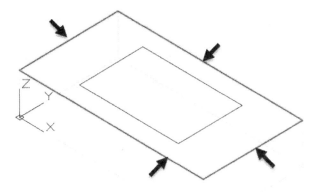

30. Press ENTER.
31. Select the **Yes** option from the command line.
32. Select the **Join** option from the command line.
33. Press ENTER.
34. Press ESC.
35. On the ribbon, click **Home** tab > **Selection** panel > **Filter** drop-down > **No Filter**.
36. On the ribbon, click **Home** tab > **Modify** panel > **3D Move**.
37. Select the newly created region and press ENTER.

39. Move upward.
40. Type 3 and press ENTER.
41. On the ribbon, click the **Home** tab > **Modeling** panel > **Solids** drop-down > **Loft**.
42. Select the large rectangle.
43. Select the small rectangle and press ENTER.

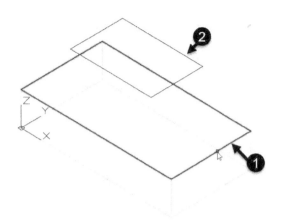

44. Select the **Settings** option from the command line.
45. On the **Loft Settings** dialog, select the **Draft angles** option.
46. Type **90** in the **Start angle** and **End angle** boxes.
47. Type **0.1** in the **Start magnitude** box.
48. Click **OK**.

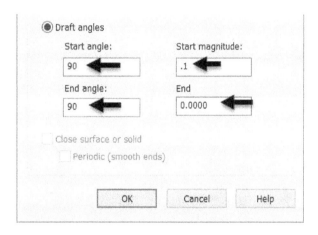

49. Change the **Visual Style** to **Shades of Gray**.

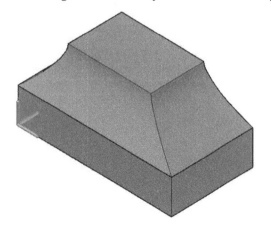

50. On the ribbon, click **Home** tab > **Solid Editing** panel > **Solid, Union**.
51. Create a selection window across all the elements of the model.
52. Press ENTER to combine all the objects.
53. Activate the **Orthomode** icon on the status bar.
54. On the ribbon, click **Home** tab > **Modeling** panel > **Primitives** drop-down > **Cylinder**.
55. Select the **Elliptical** option from the command line.
56. Select the **Center** option from the command line.
57. Select the midpoint of the front horizontal edge, as shown.

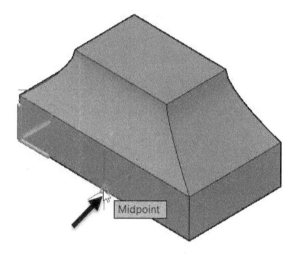

58. Type .9 and press ENTER.
59. Type 3 and press ENTER.
60. Move the pointer upward and click outside the model.

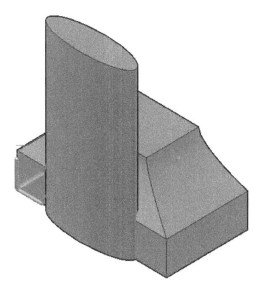

61. Select the elliptical cylinder.
62. On the ribbon, click **Home** tab > **Modify** panel > **3D Mirror**.
63. Select the **ZX** option from the command line. The orientation of the mirror plane is defined
64. Select the midpoint of the bottom-right edge, as shown. The location of the mirror plane is defined.

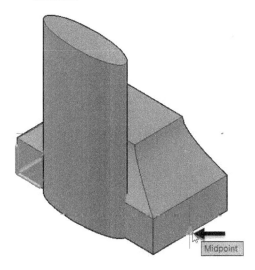

65. Select the **No** option from the command line.

66. On the ribbon, click the **Home** tab > **Modeling** panel > **Solid Editing** drop-down > **Solid, Subtract**.
67. Click on the main body
68. Press ENTER.
69. Select the two elliptical cylinders.
70. Press ENTER.

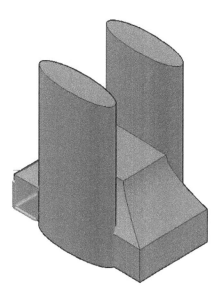

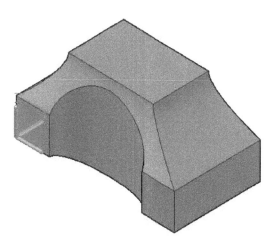

71. Change the **Visual Style** to **2D Wireframe**.
72. On the ribbon, click **Home** tab > **Selection** panel > **Filter** drop-down > **Edge**.
73. On the ribbon, click **Home** tab > **Solid Editing** panel > **Edge** drop-down > **Extract Edges**.
74. Select the edges of the model, as shown.

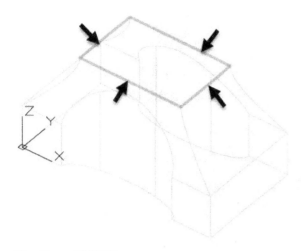

75. Press ENTER.
76. On the ribbon, click the **Home** tab > **Modify** panel > **Offset**.
77. Type 5.475, and press ENTER.
78. Select the extracted left edge.
79. Move the pointer toward the left and click.

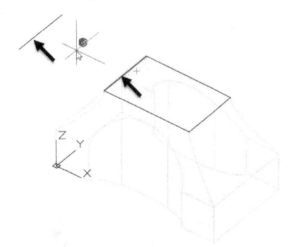

80. Select the extracted right edge.
81. Move the pointer toward the right and click.

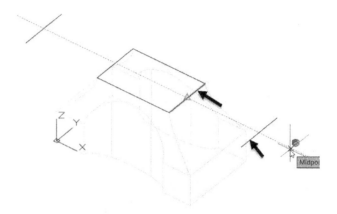

82. Press ESC.
83. Type L in the command line and press ENTER.
84. Select the front endpoints of the two offset lines.

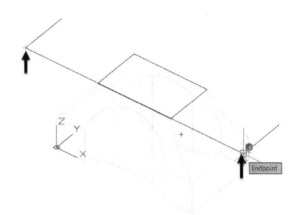

85. Press ENTER twice.
86. Select the back endpoints of the two offset lines.

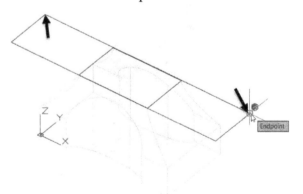

87. Press ESC.
88. On the **Home** tab of the ribbon, expand the **Modify** panel and click the **Edit Polyline** icon.
89. Select the **Multiple** option from the command line.

90. Select the offset and newly created lines, as shown.

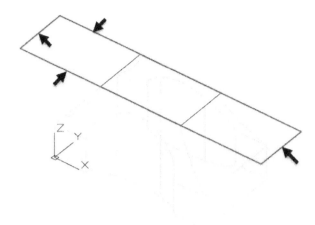

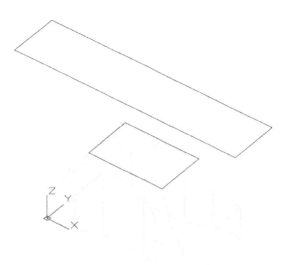

91. Press ENTER.
92. Select the **Yes** option from the command line.
93. Select the **Join** option from the command line.
94. Press ENTER.
95. Press ESC.
96. On the ribbon, click **Home** tab > **Modify** panel > **3D Move**.
97. Select the newly created polyline and press ENTER.
98. Select the Z-axis of the move gizmo.

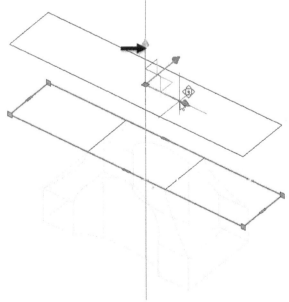

99. Move upward.
100. Type 4.8 and press ENTER.

101. On the **Home** tab of the ribbon, expand the **Modify** panel and click the **Edit Polyline** icon.
102. Select the **Multiple** option from the command line.
103. Select the extracted lines, as shown.

104. Press ENTER.
105. Select the **Yes** option from the command line.
106. Select the **Join** option from the command line.
107. Press ENTER.
108. On the ribbon, click the **Home** tab > **Modeling** panel > **Solids** drop-down > **Loft**.
109. Select the small rectangle.
110. Select the large rectangle and press ENTER.

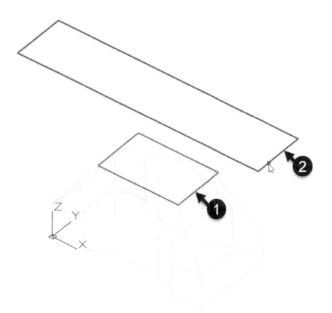

111. Select the **Settings** option from the command line.
112. On the **Loft Settings** dialog, select the **Normal to** option.
113. Select the **Start cross section** option from the drop-down.
114. Click **OK**.
115. Change the **Visual Style** to **Shades of Gray**.

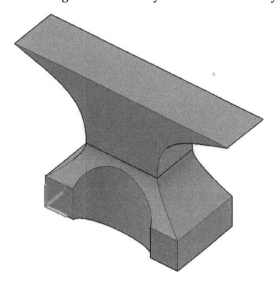

116. On the ribbon, click **Home > Modeling > Primitives** drop-down > **Box**.
117. Select the corner point of the loft solid, as shown.

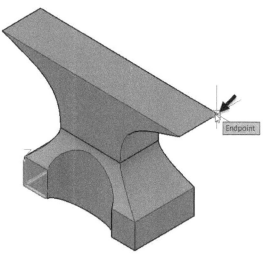

118. Select the **Length** option from the command line.
119. Select the other corner point of the loft solid, as shown; the length of the box is defined.

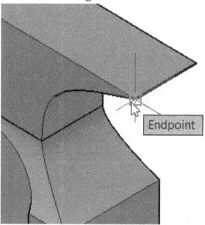

120. Move the pointer toward left and type 5.

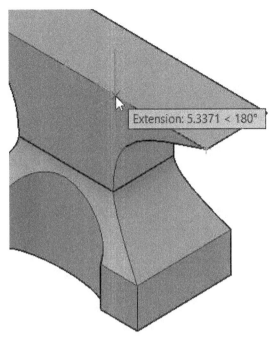

121. Move the pointer downward and select the lower corner point of the loft, as shown.

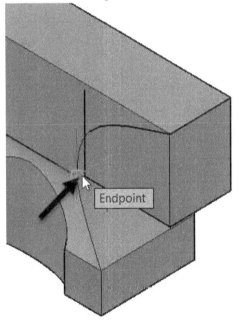

122. On the ribbon, click **Home** tab > **Solid Editing** panel > **Solid, Subtract**.
123. Select the lofted solid and press ENTER.
124. Select the box and press ENTER.

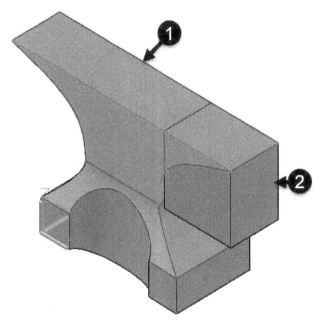

Creating a Loft with Guide Curves

1. Change the **Visual Style** to **2D Wireframe**.
2. On the ribbon, click **Home** tab > **Coordinates** panel > **View** drop-down > **Face**.

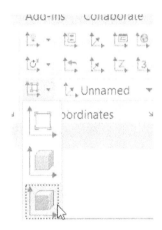

3. Click on the right-side face of the model.

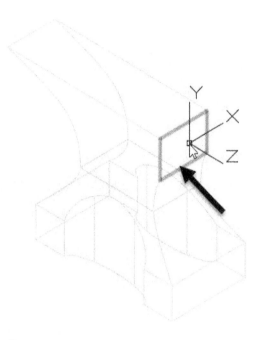

4. Press ENTER to accept the selection. The UCS is placed on the selected face.

5. On the **Home** tab of the ribbon, click **Draw > Ellipse** drop-down > **Axis, End**.

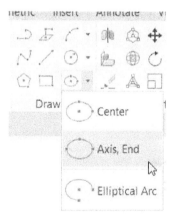

6. Select the midpoint of the left vertical edge, as shown.

7. Select the midpoint of the right vertical edge, as shown.

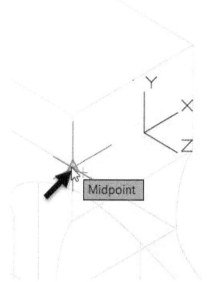

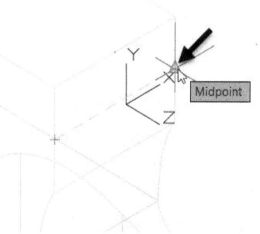

8. Select the midpoint of the lower horizontal edge, as shown.

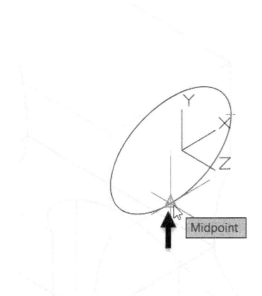

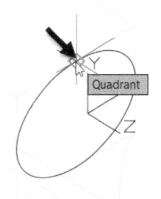

12. Move the pointer vertically downward.
13. Type 0.5 and press ENTER.

9. On the **Home** tab of the ribbon, click **Modeling > Primitives** drop-down > **Sphere**.

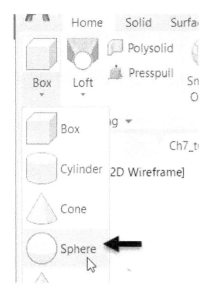

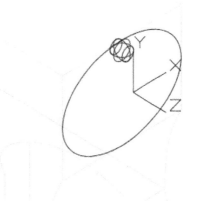

14. On the ribbon, click **Home** tab > **Modify** panel > **3D Move**.
15. Select the newly created sphere and press ENTER.
16. Click on the X-axis (Red arrow) of the move gizmo.

10. Select the **2P** option from the command line.
11. Select the top quadrant point of the ellipse.

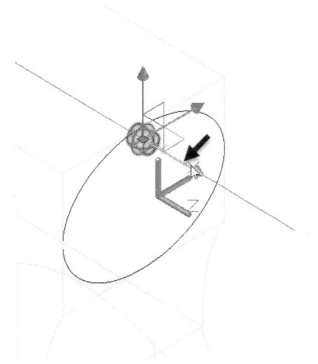

17. Move along the selected axis.

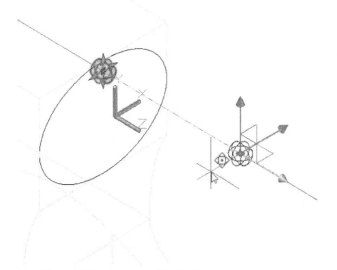

18. Type 7.72 and press ENTER.

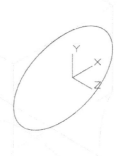

19. On the **Home** tab of the ribbon, click **Draw > Circle** drop-down > **Circle, Diameter**.
20. Select the centerpoint of the sphere.

21. Move the pointer outward.

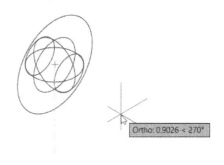

```
ircle: .5
```

22. Type .5 and press ENTER.
23. On the **Home** tab of the ribbon, click **Coordinates > Z-Axis Vector**.
24. Select the midpoint of the lower horizontal edge, as shown.

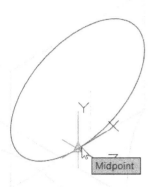

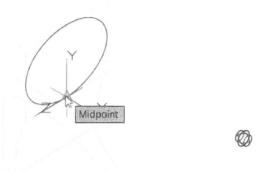

28. Move the pointer backward.

25. Move the pointer toward the left and select the corner point, as shown.

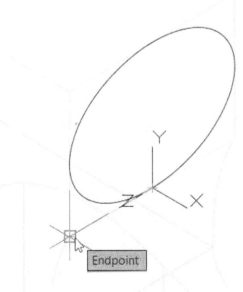

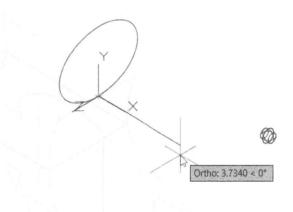

29. Type 4.11 and press ENTER.
30. Move the pointer upward.

26. On the **Home** tab of the ribbon, click **Draw** panel **> Polyline**.
27. Select the origin point of the UCS.

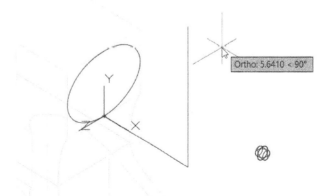

31. Type 0.9 and press ENTER.

32. Deactivate the Orthomode icon on the status bar.
33. Place the pointer on the centerpoint of the sphere.
34. Move the pointer downward; the trace line appears.
35. Select the intersection point between the trace line and circle, as shown.

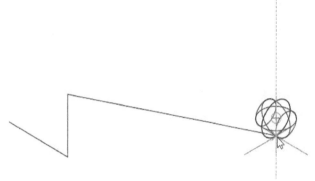

36. Press ESC.
37. On the **Home** tab of the ribbon, click **Draw** panel > **Spline**

38. Select the **Method** option from the command line.
39. Select the **Fit** option from the command line.
40. Select the points of the polyline, as shown.

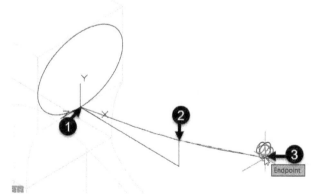

41. Right-click and select Enter.
42. Select the polyline and press DELETE.

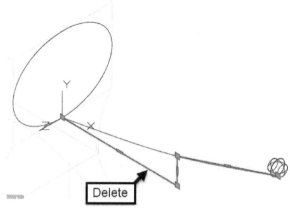

43. On the **Home** tab of the ribbon, click **Draw** panel > **Line**.
44. Select the top quadrant point of the ellipse.
45. Place the pointer on the centerpoint of the sphere.
46. Move the pointer upward.
47. Select the intersection point between the trace line and the circle.

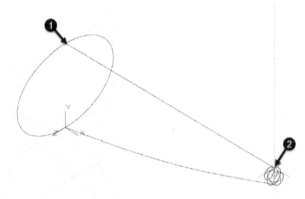

48. Press ESC.
49. On the **Home** tab of the ribbon, click **Modeling** > **Solids** drop-down > **Loft**.
50. Select the ellipse and circle. Next, press ENTER.

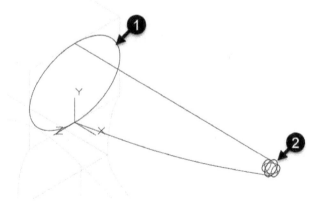

51. Select the **Guides** option.
52. Select the spline and line. Next, press ENTER.

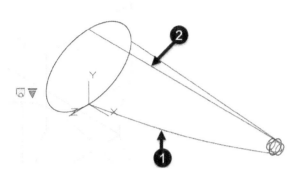

53. Change the **Visual Style** to **Shades of Gray**.

Extruding the Face

1. On the **Home** tab of the ribbon, click **Solid Editing > Face Editing** drop-down **> Extrude Faces**.

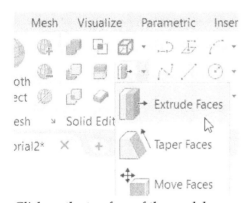

2. Click on the top face of the model.
3. Press ENTER.

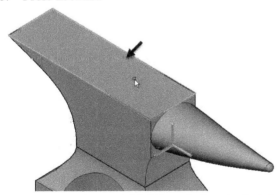

4. Type 1 and press ENTER to define the height of the extrusion.
5. Press ENTER to define 0 as the taper angle.

6. Press ESC.

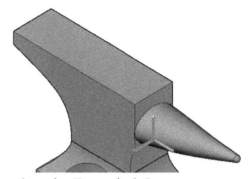

Creating the Extruded Cuts

1. On the **Home** tab of the ribbon, click **Draw** panel **> Rectangle**.
2. Select the corner point of the model, as shown.

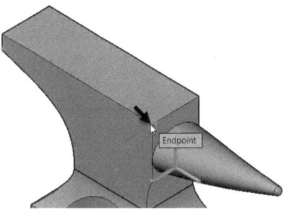

3. Select the **Dimensions** option from the command line.
4. Type 0.5 and press ENTER.
5. Type 0.25 and press ENTER.
6. Move the pointer toward left and downward, and then click.

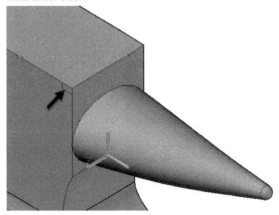

7. On the **Home** tab of the ribbon, click **Modeling > Solid** drop-down **> Extrude**.

8. Click in the region enclosed by the rectangle.

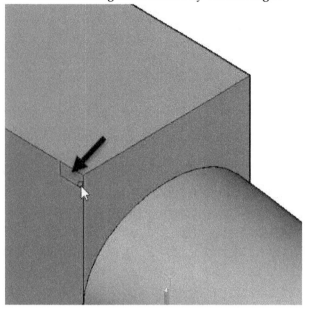

9. Move the pointer toward the right and click outside the model.

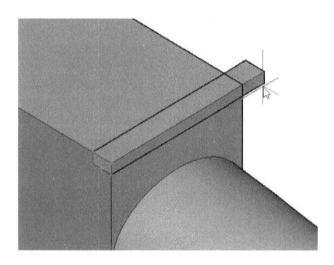

10. On the ribbon, click the **Home** tab > **Modeling** panel > **Solid Editing** drop-down > **Solid, Subtract**.
11. Click on the loft body.
12. Press ENTER.
13. Select the extrusion.
14. Press ENTER.

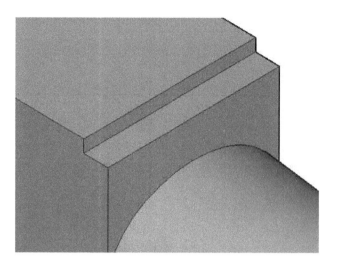

15. On the ribbon, click **Home** tab > **Coordinates** panel > **View** drop-down > **Face**.

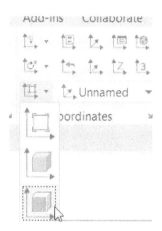

16. Click on the top face of the model.

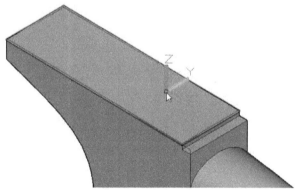

17. Press ENTER to accept the selection. The UCS is placed on the selected face.
18. Activate the Orthomode icon on the status bar.
19. On the ribbon, click **Home** > **Modeling** > **Primitives** drop-down > **Box**.

20. Select the corner point of the extrusion, as shown.

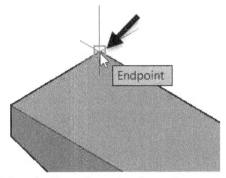

21. Select the **Length** option from the command line.
22. Type **.87**, and press ENTER.
23. Type **.87**, and press ENTER.
24. Move the pointer downward and click outside the model.

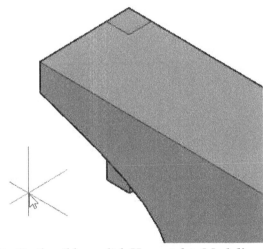

25. On the ribbon, click **Home** tab > **Modeling** panel > **Primitives** drop-down > **Cylinder**.
26. Select the corner point of the model, as shown.

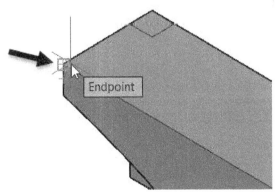

27. Type .25, and press ENTER.

28. Move the pointer downward and click outside the model.

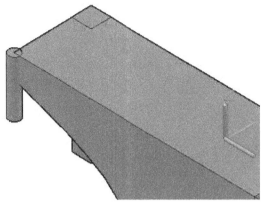

29. Select the cylinder.
30. Click on the X-axis of the move gizmo.
31. Move the pointer along the selected axis.

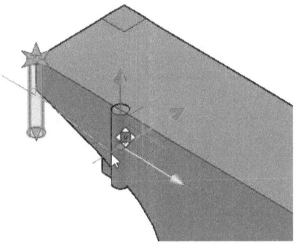

32. Type 1.2 and press ENTER.
33. Click on the Y-axis of the move gizmo.
34. Move the pointer along the selected axis.

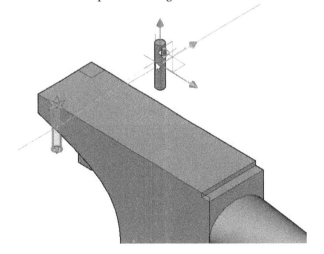

35. Type 1 and press ENTER.
36. Press ESC.
37. Select the box.
38. Click on the X-axis of the move gizmo.
39. Move the pointer along the selected axis.

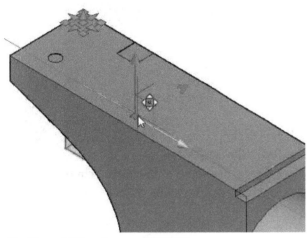

40. Type 1.87 and press ENTER.
41. Click on the Y-axis of the move gizmo.
42. Move the pointer toward left along the selected axis.

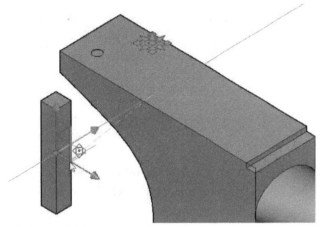

43. Type 1.34 and press ENTER.
44. Press ESC.

45. On the ribbon, click the **Home** tab > **Modeling** panel > **Solid Editing** drop-down > **Solid, Subtract**.
46. Click on the loft body.

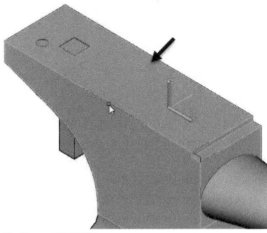

47. Press ENTER.
48. Select the cylinder and box.
49. Press ENTER.

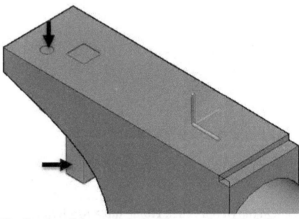

50. On the ribbon, click the **Home** tab > **Modeling** panel > **Solid Editing** drop-down > **Solid, Union**.
51. Create a selection window across all the objects.
52. Press ENTER.
53. Save and close the file.

Exercises

Exercise 1

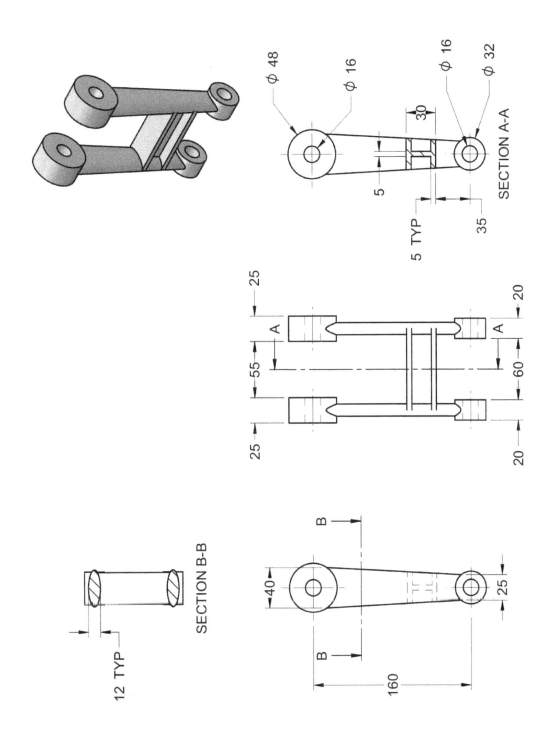

Chapter 7: Additional Features

Tutorial 1 (Millimetres)

In this example, you create the part shown next.

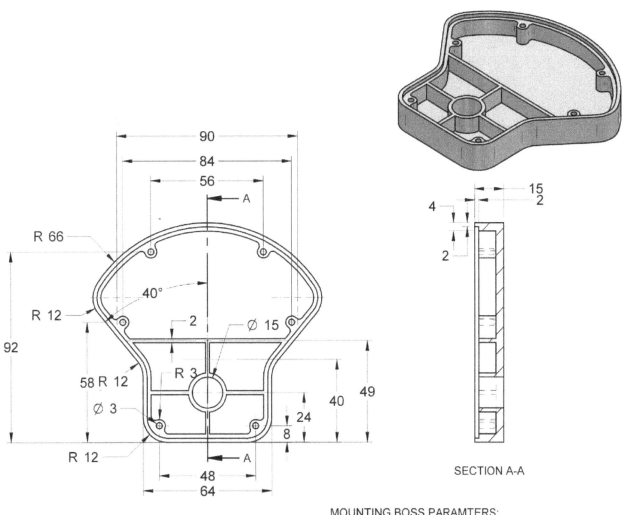

SECTION A-A

MOUNTING BOSS PARAMTERS:
 DIAMETER = 6 mm
 HOLE DIAMETER = 3 mm
 HOLE DEPTH = 8 mm

FILLET MOUNTING BOSS CORNER 2 mm

Creating a New File

1. Click the **New** icon on the Quick Access Toolbar.

2. Select the **acadiso3D** template from the **Select template** dialog.
3. Click the **Open** button.
4. Deactivate the **GRIDMODE** icon on the status bar.

Creating a Parametric Sketch

1. On the ribbon, click **Visualize** tab > **Named Views** panel > **Restore View** drop-down > **Top**.

2. Click the **Customization** button on the bottom right corner of the window.

3. Select the **Infer Constraints** option.

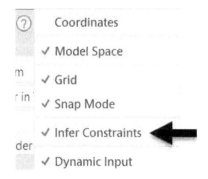

4. Click the **Infer Constraints** icon on the status bar.

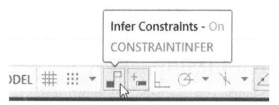

5. On the ribbon, click **Home** tab > **Draw** panel > **Line**.
6. Click in the graphics window to specify the start point of the line.
7. Activate the **Orthomode** icon on the status bar.
8. Move the pointer toward the right.
9. Type 64 and press ENTER.
10. Move the pointer upward.
11. Type 40 and press ENTER.
12. Deactivate the **Orthomode** icon on the status bar.
13. Move the diagonally toward the right and click.

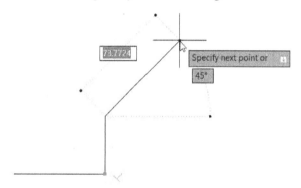

Notice that the infer constraints are created between the endpoints of the lines.

14. On the **Home** tab of the ribbon, expand the **Modify** panel and click the **Mirror** icon.
15. Select the vertical and inclined lines.
16. Press ENTER.
17. Select the midpoint of the horizontal line.
18. Activate the **Orthomode** icon on the status bar.
19. Move the pointer upward and click to specify the mirror line.

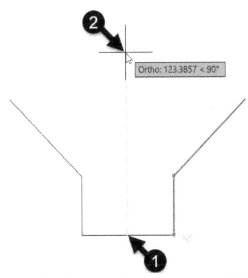

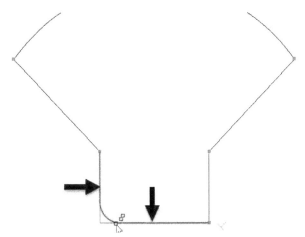

31. Press ENTER to activate the **Fillet** command.
32. Select the right vertical line and the horizontal line.
33. Press ENTER to activate the **Fillet** command.
34. Select the left vertical and left inclined lines.
35. Press ENTER to activate the **Fillet** command.
36. Select the right vertical and right inclined lines.
37. Press ENTER to activate the **Fillet** command.
38. Select the left inclined line and the arc.
39. Press ENTER to activate the **Fillet** command.
40. Select the right inclined line and the arc.

20. Select the **No** option from the command line.
21. Deactivate the **Orthomode** icon on the status bar.
22. On the ribbon, click **Home** tab > **Draw** panel > **Arc** drop-down > **3 Point**.
23. Select the endpoint of the left inclined line, as shown.
24. Move the pointer diagonally toward the right.
25. Click to specify the second point.
26. Select the endpoint of the right inclined line.

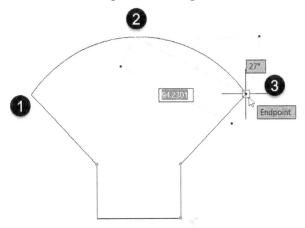

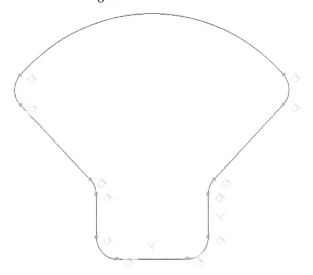

27. On the ribbon, click **Home** tab > **Modify** panel > **Fillet** drop-down > **Fillet**.
28. Select the **Radius** option from the command line.
29. Type **12** in the command line and press ENTER.
30. Select the left vertical line and the horizontal line.

41. On the ribbon, click **Parametric** tab > **Geometric** panel > **Equal**.

42. Select the two vertical lines.

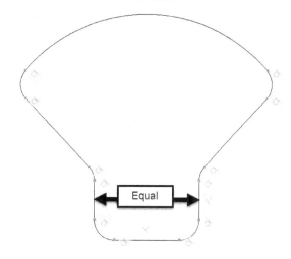

43. Press ENTER to activate the **Equal** command.
44. Select the two inclined lines to make them equal in length.

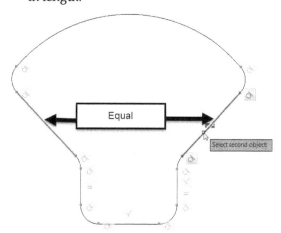

45. Press ENTER to activate the **Equal** command.
46. Select the **Multiple** option from the command line.
47. Select all the fillets and press ENTER.

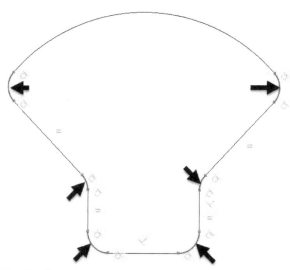

48. On the ribbon, click **Parametric** tab > **Geometric** panel > **Vertical**.

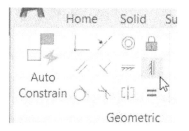

49. Select the left vertical line.

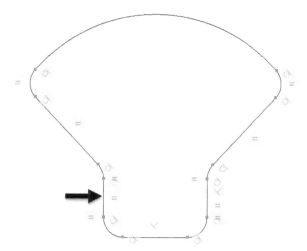

50. On the ribbon, click **Parametric** tab > **Geometric** panel > **Horizontal**.

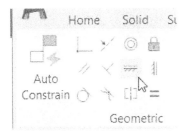

51. Select the horizontal line.

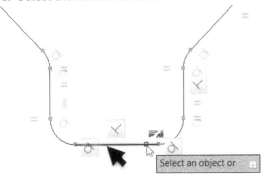

52. On the ribbon, click **Parametric** tab > **Dimensional** panel > **Linear**.

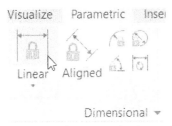

53. Select the lower endpoints of the two vertical lines.
54. Move the pointer downward and click.

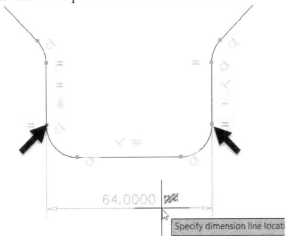

55. Type 64 and press ENTER.

56. On the ribbon, click **Parametric** tab > **Dimensional** panel > **Linear** drop-down > **Horizontal**.

57. Press and hold the Shift key and right-click.
58. Select **Center** from the **Object Snaps** menu.
59. Click on the left fillet, as shown. The center point of the fillet is selected.

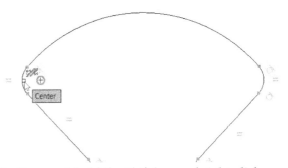

60. Press and hold the Shift key and right-click.
61. Select **Center** from the **Object Snaps** menu.
62. Click on the right fillet, as shown. The center point of the fillet is selected.

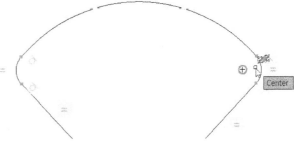

63. Move the pointer upward and click to create the horizontal constraint.
64. Click in the graphics window.

65. On the ribbon, click **Parametric** tab > **Dimensional** panel > **Angular**.

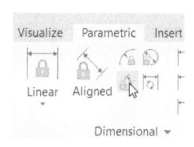

66. Select the left vertical and inclined lines, as shown.
67. Move the pointer toward the left and click.
68. Type 40 and press ENTER.

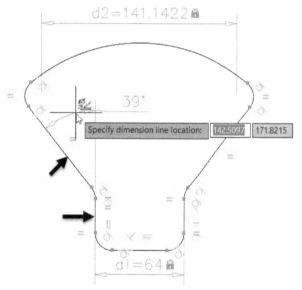

69. On the ribbon, click **Home** tab > **Draw** panel > **Line**.
70. Activate the **Orthomode** icon on the status bar.
71. Select the midpoint of the horizontal line, as shown.
72. Move the pointer vertically upward and click.

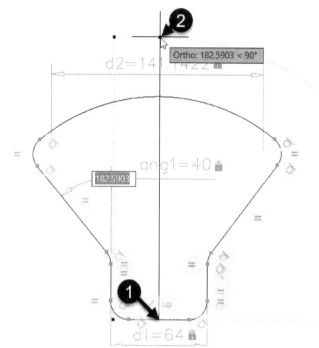

73. On the ribbon, click **Parametric** tab > **Geometric** panel > **Symmetric**.

74. Select the left inclined line to define the first object.
75. Select the right inclined line to define the second object.
76. Select the vertical line to define the symmetry line.

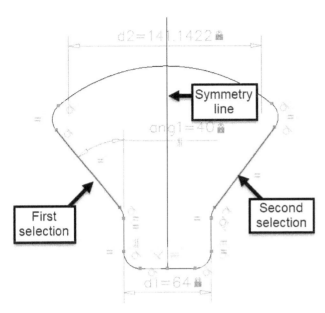

77. On the ribbon, click **Parametric** tab > **Dimensional** panel > **Radius**.

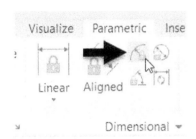

78. Select the large arc.
79. Move the pointer and click.
80. Click in the graphics window

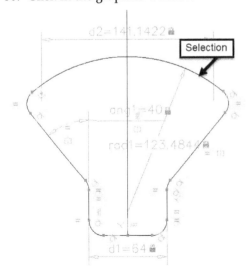

81. On the ribbon, click **Parametric** tab > **Dimensional** panel > **Linear** drop-down > **Vertical**.

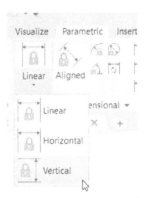

82. Press and hold the Shift key and right-click.
83. Select **Center** from the **Object Snaps** menu.
84. Click on the large arc, as shown. The center point of the arc is selected.

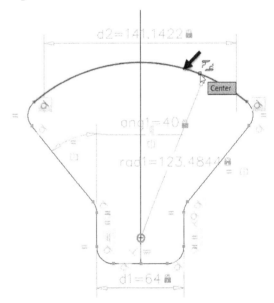

85. Select the right point of the horizontal line, as shown.

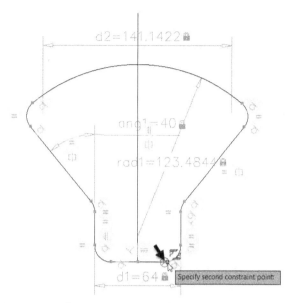

86. Move the pointer toward the right and click.
87. Type 40 and press ENTER.

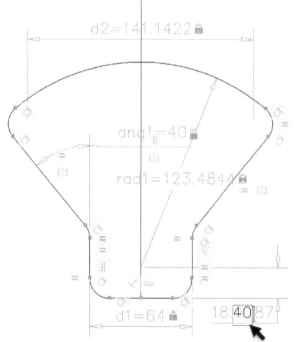

88. On the ribbon, click **Parametric** tab >
    Dimensional panel > **Radius**.
89. Select any one of the fillets.
90. Place the dimensional constraint.
91. Type **12** and press ENTER.

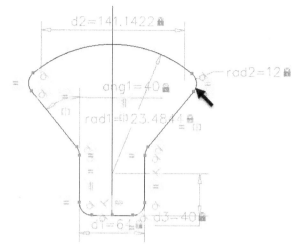

92. Double-click on the horizontal constraint, as
    shown.
93. Type 90 and press ENTER.

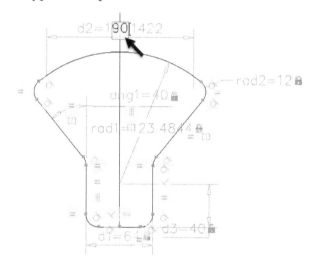

94. Select the radius constraint, as shown.
95. Type 66 and press ENTER.

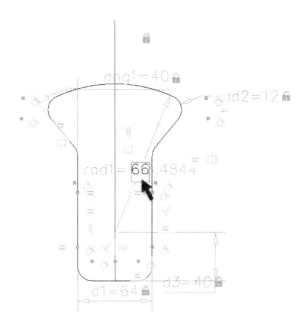

Extruding the Sketch

1. On the **Home** tab of the ribbon, expand the **Draw** panel and click the **Region**.

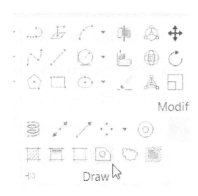

2. Create a selection window from left to right across all the elements of the sketch.

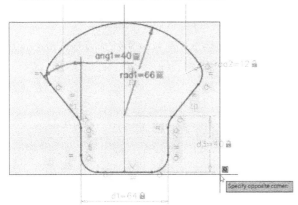

3. Press ENTER to convert the closed sketch into a region.

4. Change the view orientation to SE Isometric.
5. On the ribbon, click **Home** tab > **Modeling** panel > **Solids** drop-down > **Extrude**.
6. Select the region and press ENTER.
7. Move the pointer upward.
8. Type 14 and press ENTER.
9. Select the vertical line and press DELETE.

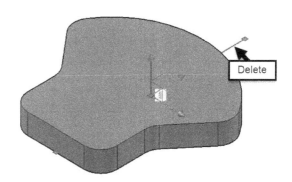

10. On the ribbon, click the **Solid** tab > **Solid Editing** panel > **Offset Edge**.

11. Click on the top face of the solid body.
12. Select the **Distance** option from the command line.

13. Type **2** in the command line and press ENTER.
14. Click inside the solid body to specify the direction of the offset edge.

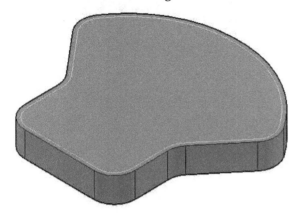

15. On the ribbon, click the **Solid** tab > **Solid Editing** panel > **Shell**.
16. Select the solid body.
17. Click on the top face of the body.
18. Press ENTER.
19. Type **4** in the command line and press ENTER.
20. Press ENTER twice to exit the command.

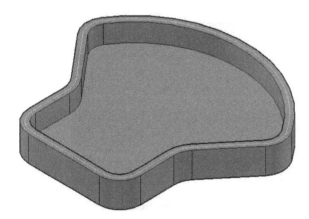

Adding a Lip to the model

1. Select the offset edge.

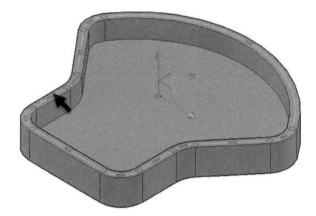

2. On the ribbon, click **Home** tab > **Modeling** panel > **Solids** drop-down > **Extrude**.
3. Move the pointer downward.
4. Type 2 and press ENTER.

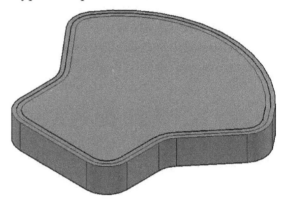

5. On the ribbon, click the **Home** tab > **Modeling** panel > **Solid Editing** drop-down > **Solid, Subtract**.
6. Click on the side face of the model.
7. Press ENTER.
8. Click on the top face of the extruded body.

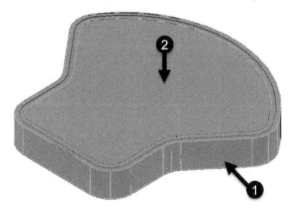

9. Press ENTER.

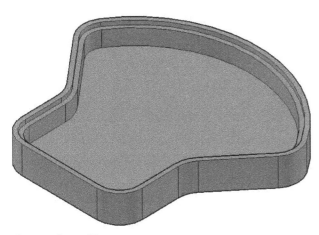

Creating Bosses

1. On the ribbon, click **Home** tab > **Coordinates** panel > **Origin**.

2. Select the midpoint of the horizontal edge, as shown.

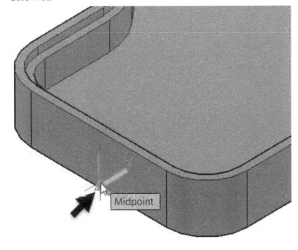

10. Click the Top face of the ViewCube.

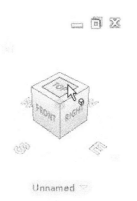

11. On the ribbon, click **Home** tab > **Draw** panel > **Circle** drop-down > **Circle, Radius**.
12. Create two circles, as shown.

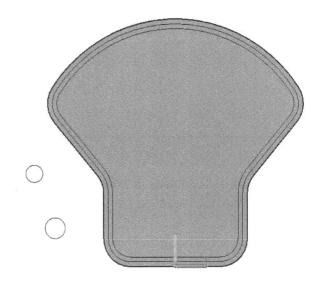

13. On the ribbon, click **Parametric** tab > **Geometric** panel > **Concentric**.

14. Select the two circles to make them concentric.
15. On the ribbon, click **Parametric** tab > **Dimensional** panel > **Diameter**.

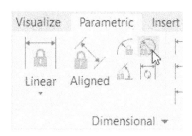

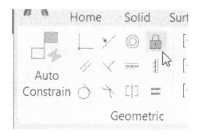

16. Select any one of the circles.
17. Position the dimensional constraint.
18. Type 6 and press ENTER.
19. On the ribbon, click **Parametric** tab > **Dimensional** panel > **Diameter**.
20. Select the other circle.
21. Position the dimensional constraint.
22. Type 3 and press ENTER.

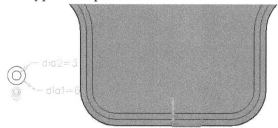

23. Change the **Visual Style** to **2D Wireframe**.
24. On the ribbon, click **Home** tab > **Draw** panel > **Line**.
25. Create a vertical line passing through the origin.

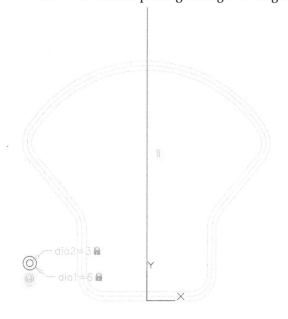

26. On the ribbon, click **Parametric** tab > **Geometric** panel > **Fix**.

27. Select the lower endpoint of the newly created vertical line.
28. On the ribbon, click **Parametric** tab > **Dimensional** panel > **Linear**.
29. Select the centerpoint of the circle.
30. Select the lower endpoint of the vertical line.
31. Move the pointer downward and click.

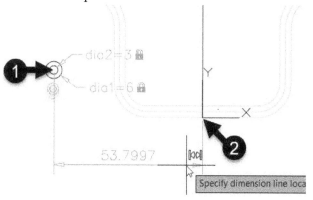

32. Type 24 and press ENTER.

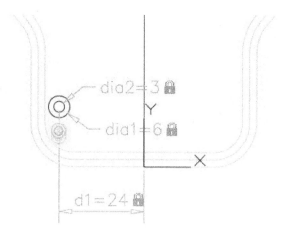

33. On the ribbon, click **Parametric** tab > **Dimensional** panel > **Linear**.
34. Select the lower endpoint of the vertical line.
35. Select the centerpoint of the circle.
36. Move the pointer toward the left and click.

94

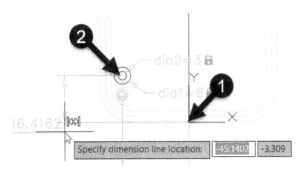

37. Type 8 and press ENTER.

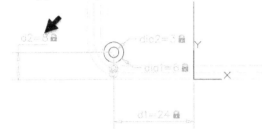

38. On the ribbon, click **Home** tab > **Modify** panel > **Copy**.
39. Select the two circles and press ENTER.
40. Click at an arbitrary point to specify the base point.
41. Move the pointer upward and click.
42. Again, move the pointer upward and click.

43. Press ESC.
44. Create the Linear dimensional constraints, as shown.

45. On the ribbon, click the **Home** tab > **Modeling** panel > **Extrude**.
46. Select all the large circles and press ENTER.

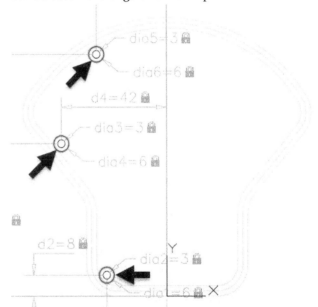

47. Type **12** and press ENTER.
48. On the ribbon, click the **Home** tab > **Modeling** panel > **Extrude**.
49. Select all the small circles and press ENTER.
50. Type 20 and press ENTER.
51. Change the view orientation to SE Isometric.
52. Change the **Visual Style** to **Shades of Gray**.

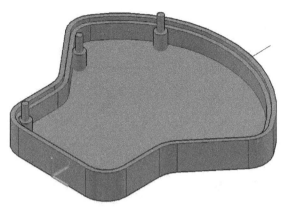

53. Select all the cylinders from the graphics window.
54. On the ribbon, click **Home** tab > **Modify** panel > **3D Mirror**.

55. Select the **YZ** option from the command line.
56. Select the origin of the UCS to define the location of the mirror plane.
57. Select the **No** option.

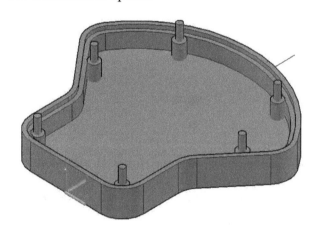

58. On the ribbon, click **Home** tab > **Solid Editing** panel > **Solid, Subtract**.
59. Select the short cylinders and press ENTER.

60. Select the tall cylinders and press ENTER.

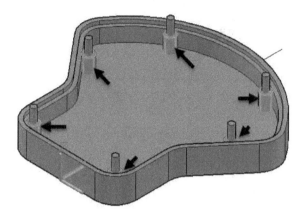

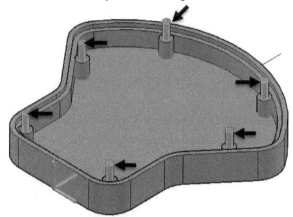

61. Select the vertical line and press DELETE.

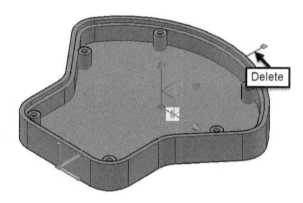

62. On the ribbon, click **Solid** tab > **Boolean** panel > **Union**.
63. Create a selection window across the entire solid body.
64. Press ENTER to combine the bosses and main body.

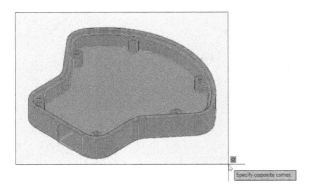

65. Change the **Visual Style** to **2D Wireframe**.
66. On the ribbon, click **Solid** tab > **Solid Editing** panel > **Fillet Edge**.
67. Select the edges between the bosses and the main body.

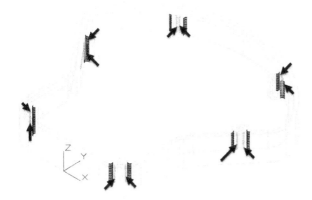

68. Select the **Radius** option from the command line.
69. Type **2** and press ENTER.
70. Press ENTER twice to fillet the edges.

Creating the Rib feature

1. On the ribbon, click **Home** tab > **Coordinates** panel > **Origin**.

2. Select the midpoint of the inner horizontal edge, as shown.

3. Click the Top face of the ViewCube.
4. On the ribbon, click **Home** tab > **Draw** panel > **Line**.
5. Select the origin point of the UCS.
6. Move the pointer upward.
7. Type 45 and press ENTER.

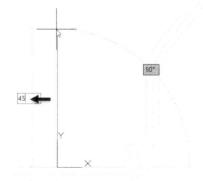

8. Move the pointer toward the right and click outside the model.

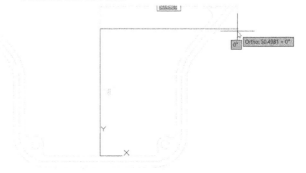

9. Press ESC.
10. On the ribbon, click **Home** tab > **Draw** panel > **Line**.
11. Select the endpoint of the vertical line.

12. Move the pointer toward the left and click outside the model.

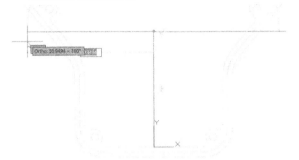

13. On the ribbon, click **Home** tab > **Selection** panel > **Filter** drop-down > **Edge**.

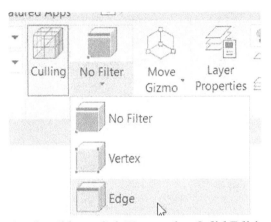

14. On the ribbon, click **Home** tab > **Solid Editing** panel > **Edge** drop-down > **Extract Edges**.

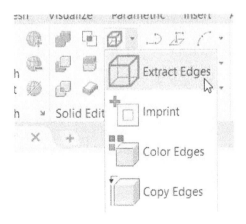

15. Select the edges of the model, as shown.

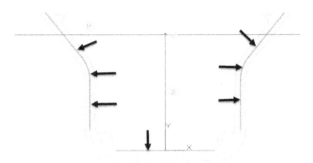

16. Press ENTER.
17. On the ribbon, click the **Home** tab > **Modify** panel > **Offset**.
18. Type 25 and press ENTER.
19. Select the horizontal line, as shown.
20. Move the pointer downward and click.

21. Likewise, offset the other horizontal line, as shown.

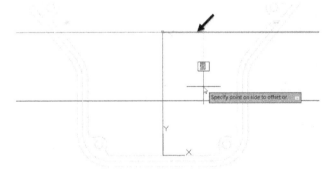

22. On the ribbon, click **Home** tab > **Draw** panel > **Circle** drop-down > **Circle, Diameter**.
23. Select the intersection point between the vertical and horizontal lines.
24. Type 17 and press ENTER.

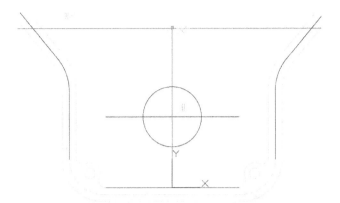

25. On the ribbon, click **Home** tab > **Modify** panel > **Trim/Extend** drop-down > **Trim**.
26. Select the portions of the horizontal line extending outside the model.
27. Select the lines inside the circle.

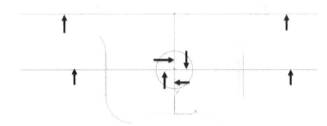

28. Press ESC.

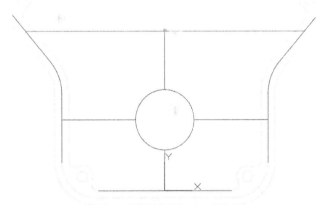

29. On the ribbon, click the **Home** tab > **Modeling** panel > **Polysolid**.

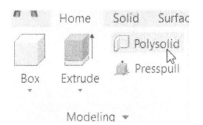

30. Select the **Height** option from the command line.
31. Type 8 and press ENTER.
32. Select the **Width** option from the command line.
33. Type 2 and press ENTER.
34. Select the **Justify** option from the command line.
35. Select the **Center** option from the command line.
36. Select the **Object** option from the command line.
37. Select the horizontal line, as shown.

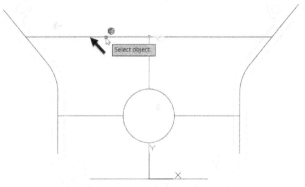

38. Press ENTER and select the **Object** option.
39. Select the next horizontal line, as shown.

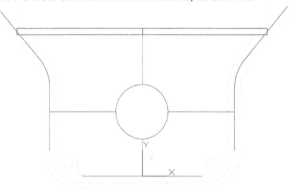

40. Likewise, use the **Polysolid** command and select the remaining lines and circle.

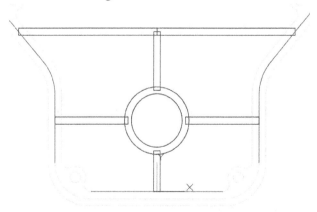

41. Change the view orientation to **SE Isometric**.
42. Change the **Visual Style** to **Shades of Gray**.

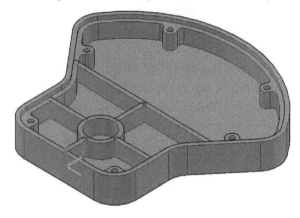

43. On the ribbon, click **Home** tab > **Solid Editing** panel > **Solid, Union**.

44. Create a selection window across all the objects of the model and press ENTER.

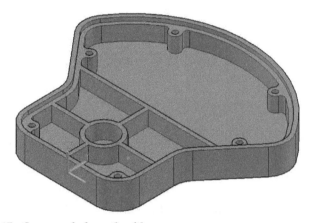

45. Save and close the file.

Tutorial 2 (Inches)

In this example, you create the part shown next.

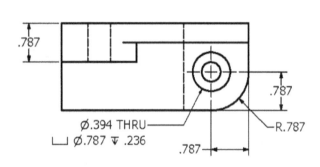

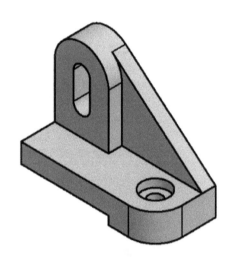

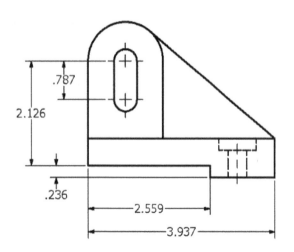

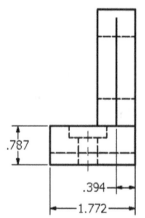

Creating a New File

1. Click the **AutoCAD 2021** icon on your desktop.
2. On the **Start** page, click **Get Started** drop-down > **acad3D.dwt**.
3. On the status bar, click **Workspace Switching** drop-down > **3D Modeling**.
4. Deactivate the **GRIDMODE** icon on the status bar.

Creating Extruded Bodies

1. On the ribbon, click **Home > Draw > Rectangle**.

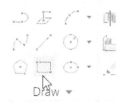

2. Type 0,0 in the command line and press ENTER. The first corner of the rectangle is defined.
3. Select **Dimensions** from the command line.
4. Type **3.937** in the command line and press ENTER. The length of the rectangle is defined.
5. Type **1.772** in the command line and press ENTER. The width of the rectangle is defined.
6. Move the pointer toward the left and click to create the rectangle.
7. On the ribbon, click **Home > Modeling > Extrude**.
8. Select the rectangle and press ENTER.
9. Move the pointer upward.
10. Type **0.787**, and press ENTER.

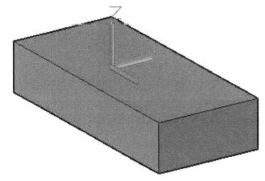

11. On the ribbon, click **Home > Coordinates > Z-Axis Vector**.

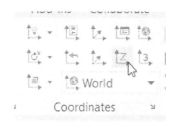

12. Select the back left corner of the extrusion.
13. Select the front left corner of the extrusion.

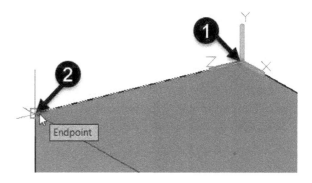

14. On the ribbon, click **Home > Modeling > Primitives** drop-down > **Box**.
15. Type 0,0 and press ENTER.
16. Move the pointer toward the right.
17. Type 1.574 and press the TAB key.
18. Type 1.575, and press ENTER.

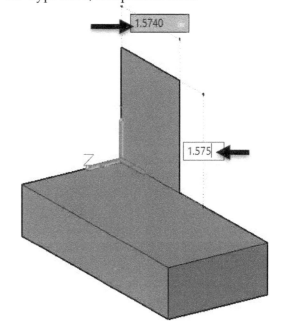

19. Move the pointer toward left.
20. Type .787 and press ENTER.

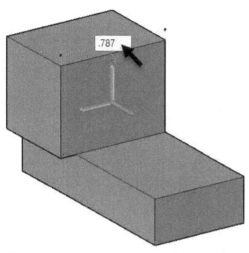

21. On the ribbon, click **Home > Modeling > Primitives** drop-down > **Cylinder**.
22. Select the midpoint of the horizontal back edge, as shown.

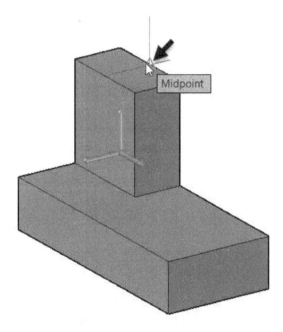

23. Move the pointer outward.
24. Select the corner point of the box, as shown.

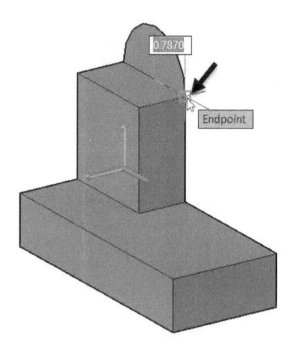

25. Move the pointer toward the left and select the corner point on the front face of the box.

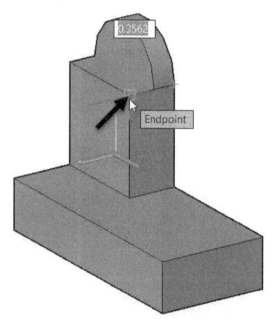

26. On the ribbon, click **Home > Solid Editing > Solid, Union**.
27. Select all the objects of the model and press ENTER.

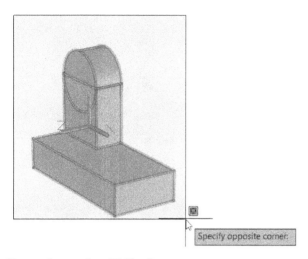

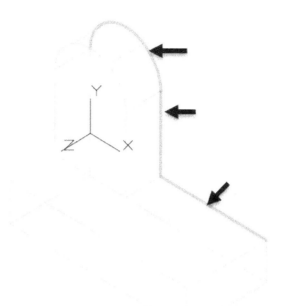

Creating the Rib feature

1. Change the **Visual Style** to **2D Wireframe**.
2. On the ribbon, click **Home** tab > **Selection** panel > **Filter** drop-down > **Edge**.

5. Press ENTER.
6. On the ribbon, click **Home** tab > **Draw** panel > **Line**.
7. Select the corner point of the model, as shown.
8. Place the pointer on the extracted arc.
9. Click when the Tangent snap appears.

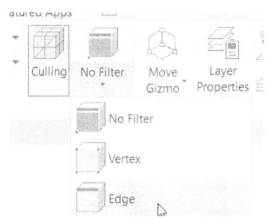

3. On the ribbon, click **Home** tab > **Solid Editing** panel > **Edge** drop-down > **Extract Edges**.

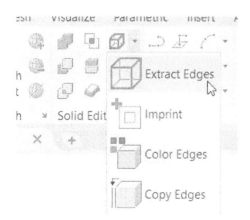

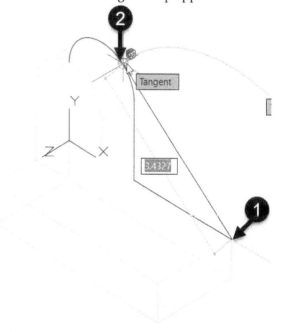

10. Press ESC.
11. On the ribbon, click **Home** tab > **Modify** panel > **Trim/Extend** drop-down > **Trim**.

4. Select the edges of the model, as shown.

12. Select the unwanted portion of the arc, as shown.

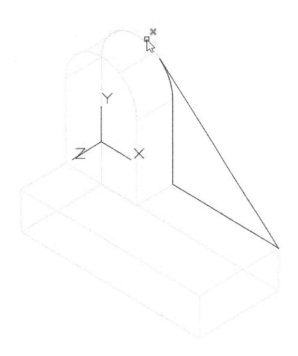

13. On the **Home** tab of the ribbon, expand the **Draw** panel and click the **Region** icon.

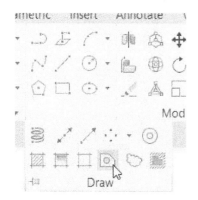

14. Select the extracted edges and the line, as shown.

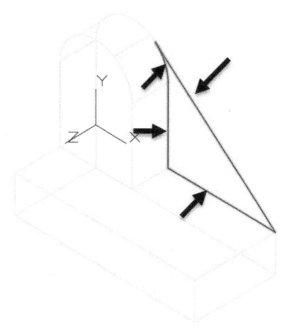

15. Press ENTER.
16. Change the **Visual Style** to **Shades of Gray**.
17. On the **Home** tab of the ribbon, click Modeling panel > **Solids** drop-down > **Extrude**.
18. Select the region and press ENTER.
19. Move the pointer toward left.
20. Type .394 and press ENTER.

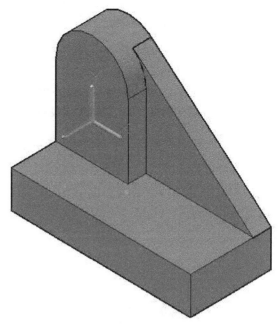

Creating the Slot

1. On the **Home** tab of the ribbon, click **Coordinates > Z-Axis Vector**.

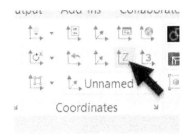

2. Select the corner point of the model, as shown.
3. Move the pointer toward the left and select the corner point, as shown. The Z-axis of the UCS is defined.

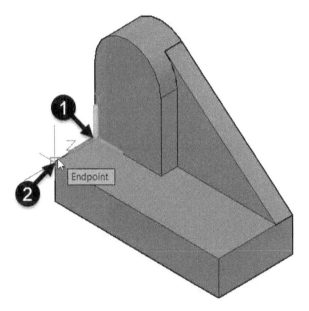

4. On the ribbon, click **Home** tab > **Modeling** panel > **Primitives** drop-down > **Cylinder**.
5. Select the centerpoint of the curved edge, as shown. The centerpoint of the cylinder is defined.

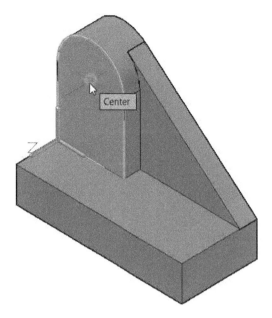

6. Type **.237** in the command line and press ENTER. The radius of the cylinder is defined.
7. Move the pointer toward the right and click outside the model.

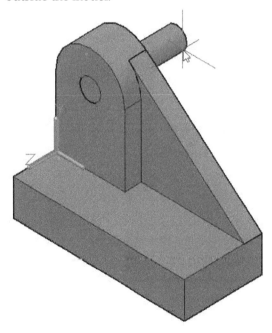

8. Select the newly created cylinder.
9. On the ribbon, click **Home** tab > **Modify** panel > **Copy**.
10. Select the center point of the cylinder to define the base point.
11. Activate the **Orthomode** icon on the Status bar.

12. Move the pointer downward.

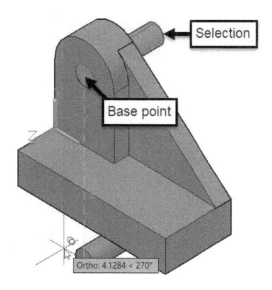

13. Type .787 and press ENTER.

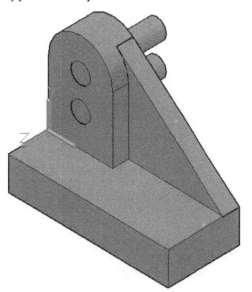

14. On the ribbon, click **Home** tab > **Modeling** panel > **Primitives** drop-down > **Box**.
15. On the status bar, click **Object Snap** drop-down > **Quadrant**.
16. Select the left quadrant point of the first cylinder.
17. Move the pointer downward.
18. Select the right quadrant point of the second cylinder.

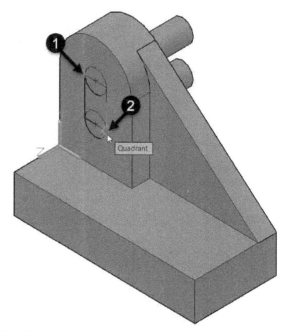

19. Move the pointer toward the right and click outside the model.

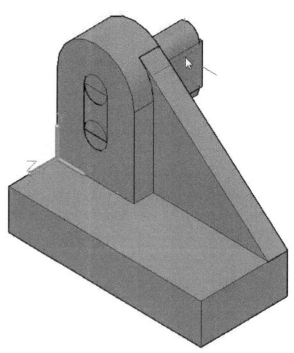

20. On the ribbon, click **Home** tab > **Solid Editing** panel > **Solid, Subtract**.
21. Select the main body and press ENTER.

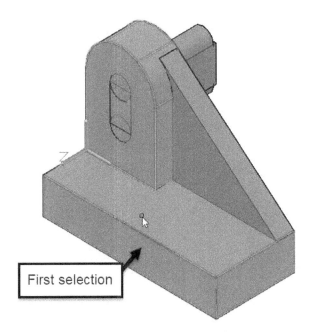

First selection

22. Select the two cylinders and the box.

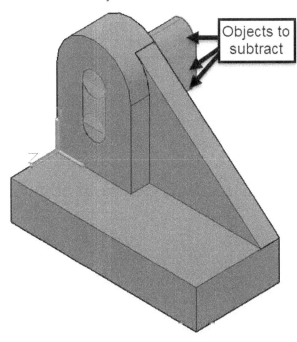

Objects to subtract

23. Press ENTER to subtract the cylinders and box from the main body.

24. On the ribbon, click **Solid** tab > **Solid Editing** panel > **Fillet Edge** drop-down > **Fillet Edge**.

25. Select the right vertical edge of the model, as shown.

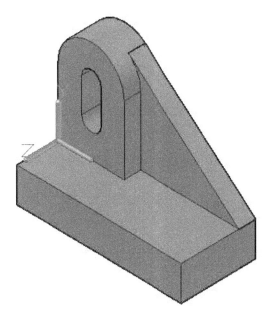

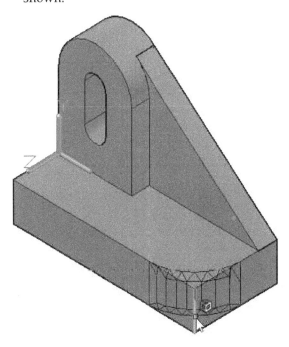

26. Select the **Radius** option from the command line.
27. Type .787 and press ENTER.
28. Press ENTER twice to create the fillet.
29. On the ribbon, click **Home** tab > **Coordinates** panel > **Z-Axis Vector**.

30. Select the centerpoint of the fillet to define the origin of the UCS.

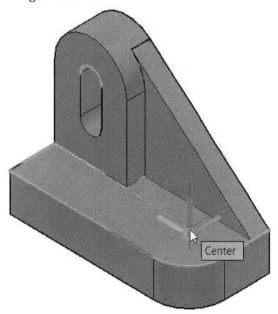

31. Move the pointer vertically downward and click.

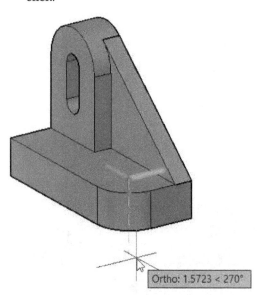

32. On the ribbon, click **Home** tab > **Modeling** panel > **Primitives** drop-down > **Cylinder**.
33. Type 0,0 and press ENTER.
34. Type .3935 and press ENTER.
35. Type 0.236, and press ENTER.

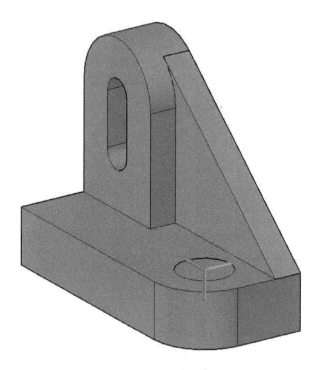

36. On the ribbon, click **Home** tab > **Modeling** panel > **Primitives** drop-down > **Cylinder**.
37. Type 0,0 and press ENTER.
38. Type .197 and press ENTER.
39. Move the pointer downward and click outside the model.

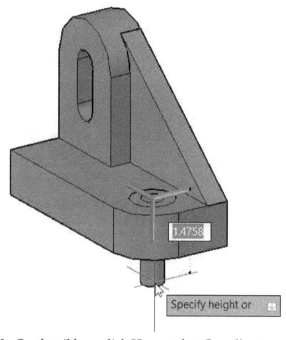

40. On the ribbon, click **Home** tab > **Coordinates** panel > **UCS, World**.

41. Deactivate the **Dynamic Input** icon on the status bar.
42. On the ribbon, click **Home** tab > **Modeling** panel > **Primitives** drop-down > **Box**.
43. Select the lower-left corner of the model.

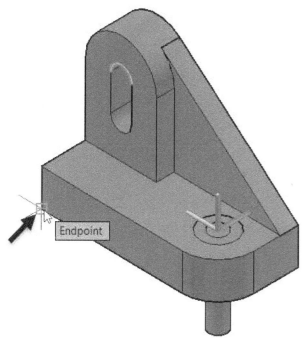

44. Select the **Length** option from the command line.
45. Type 2.559, and press ENTER.
46. Move the pointer toward the right and click outside the model.

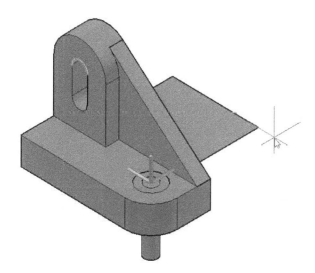

47. Type 0.236 and press ENTER to specify the height of the box.

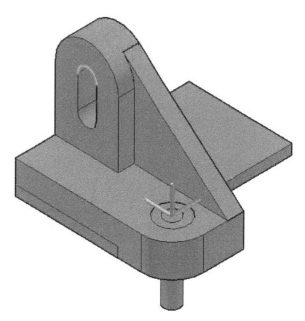

48. On the ribbon, click **Home** tab > **Solid Editing** panel > **Solid, Subtract**.
49. Select the main body and press ENTER.

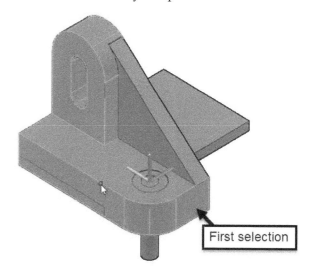

50. Select the two cylinders and the box.

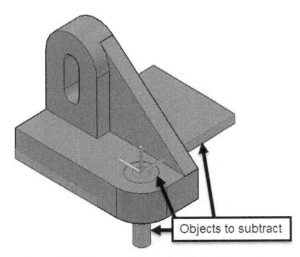

Objects to subtract

51. Press ENTER to subtract the objects from the main body.

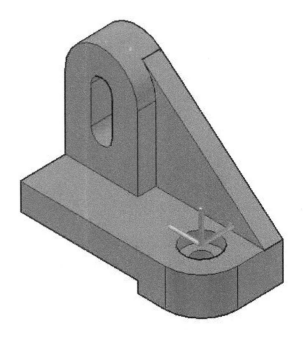

52. Save and close the file.

Exercises
Exercise 1

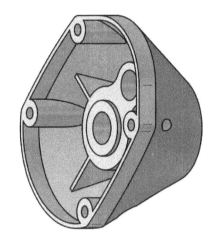

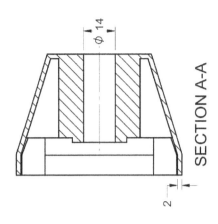

SECTION A-A

φ 14

2

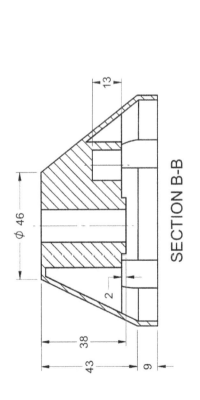

SECTION B-B

13

φ 46

2

38

43

9

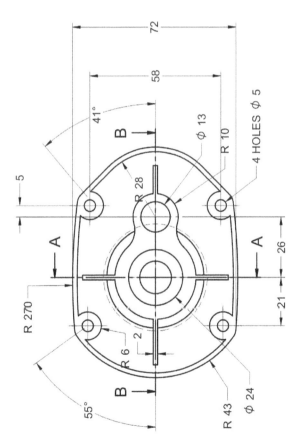

72

58

41°

5

B

φ 13

R 10

4 HOLES φ 5

R 28

A

A

26

21

R 270

R 6

2

55°

B

R 43

φ 24

Exercise 2

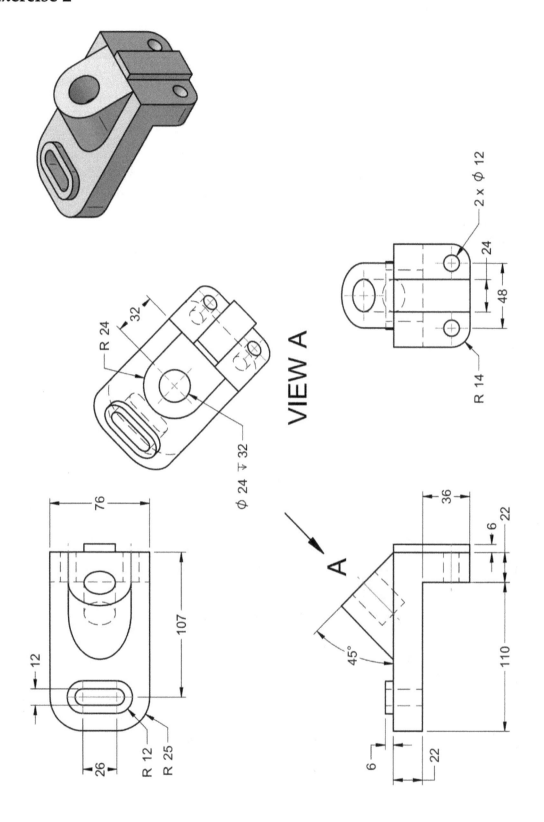

VIEW A

Exercise 3 (Inches)

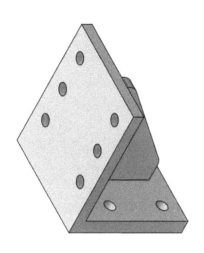

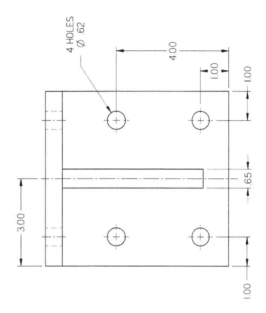

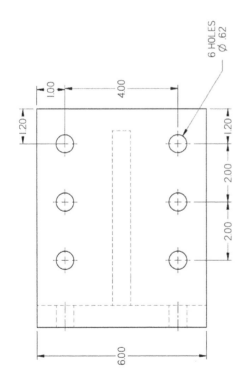

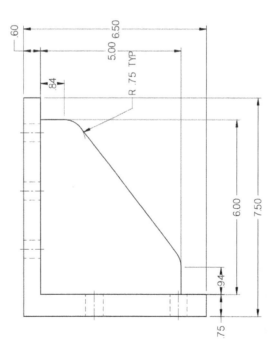

Chapter 8: Modifying Parts

Tutorial 1 (Millimetres)

In this example, you create the part shown below and then modify it using the editing tools.

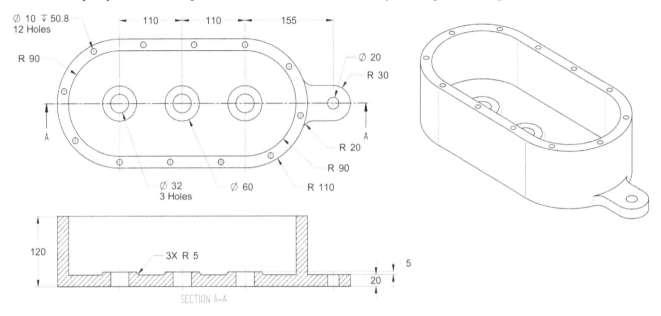

Creating a New File

1. Click the **New** icon on the Quick Access Toolbar.

2. Select the **acadiso3D** template from the **Select template** dialog.
3. Click the **Open** button.
4. Deactivate the **GRIDMODE** icon on the status bar.

Creating the Base

1. Activate the **Orthomode** icon on the status bar.
2. On the ribbon, click **Home** tab > **Modeling** panel > **Primitives** drop-down > **Box**.
3. Type 0,0 and press ENTER to specify the first corner of the box.
4. Select the **Length** option from the command line.
5. Type 220, and press ENTER.

6. Type 220, and press ENTER.
7. Type 120, and press ENTER.
8. On the ribbon, click **Home** tab > **Modeling** panel > **Primitives** drop-down > **Cylinder**.
9. Select the midpoint of the top-left edge, as shown.

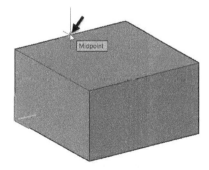

10. Select the top-left corner of the box.

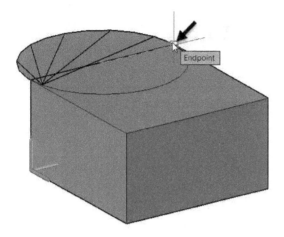

11. Move the pointer downward and select the bottom corner of the box.

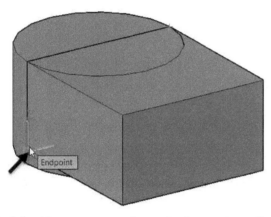

12. Likewise, create another cylinder on the right side of the box.

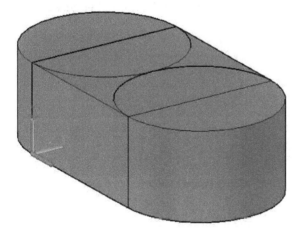

13. On the ribbon, click **Home** tab > **Solid Editing** panel > **Solid, Union**.

14. Create a selection window across all the elements of the model.
15. Press ENTER to combine all the objects.
16. On the ribbon, click the **Solid** tab > **Solid Editing** panel > **Shell**.
17. Select the solid body.
18. Click on the top face of the body.
19. Press ENTER.
20. Type **20** in the command line and press ENTER.

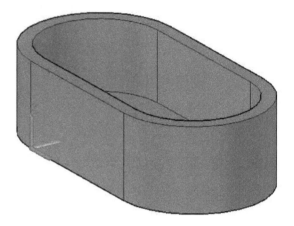

Adding Features and Holes

1. On the ribbon, click **Home** tab > **Modeling** panel > **Primitives** drop-down > **Cylinder**.
2. Select the midpoint of the curved bottom edge, as shown.

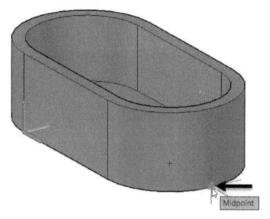

3. Type 30 and press ENTER to specify the radius.
4. Type 20 and press ENTER to specify the height.

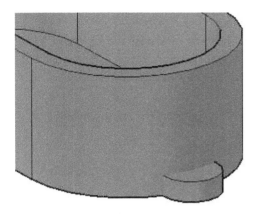

5. Select the newly created cylinder.
6. Select the X-axis of the move gizmo.

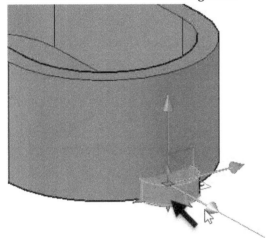

7. Move along the selected axis.

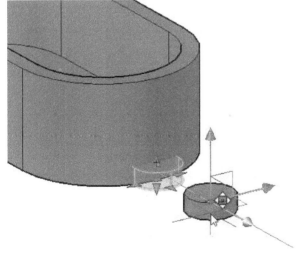

8. Type 45 and press ENTER.

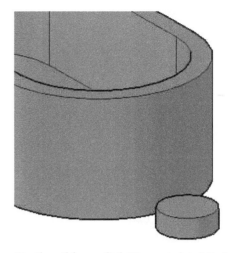

9. On the ribbon, click **Home** tab > **Modeling** panel > **Primitives** drop-down > **Box**.
10. Select the quadrant point of the lower circular edge of the cylinder.

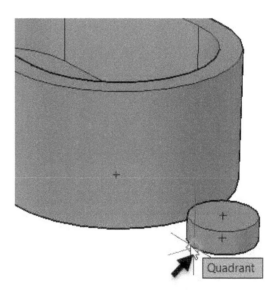

11. Select the **Length** option from the command line.
12. Type -60 and press ENTER.
13. Type -60 and press ENTER.
14. Type 20 and press ENTER.

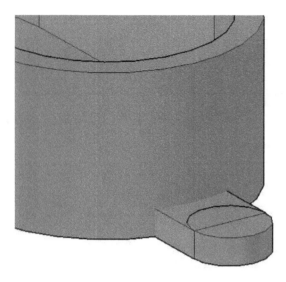

15. On the ribbon, click **Home** tab > **Coordinates** panel > **View** drop-down > **Face**.

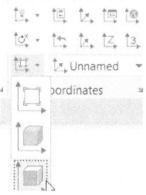

16. Click on the horizontal face of the shell.

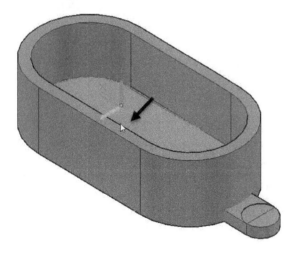

17. Press ENTER to accept the selection. The UCS is placed on the selected face.

18. On the ribbon, click **Home** tab > **Modeling** panel > **Primitives** drop-down > **Cylinder**.
19. Place the pointer on the circular edge on the left side; the centerpoint of the circular edge is displayed.
20. Select the centerpoint of the circular edge.

21. Type 30 and press ENTER.
22. Type 5 and press ENTER.

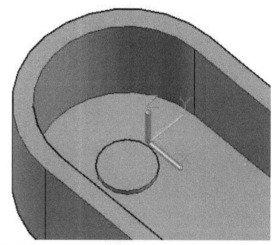

23. Select the newly created cylinder.
24. On the ribbon, click **Home** tab > **Modify** panel > **Array** drop-down > **Rectangular Array**.
25. On the **Array Creation** tab of the ribbon, type **3**, and **1** in the **Columns** and **Rows** boxes, respectively.
26. Type **110** in the **Between** box on the **Columns** panel.
27. Click the **Close Array** icon.

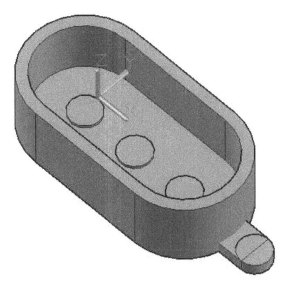

28. On the ribbon, click **Home** tab > **Solid Editing** panel > **Solid, Union**.
29. Create a selection window across all the objects of the model, and press ENTER.

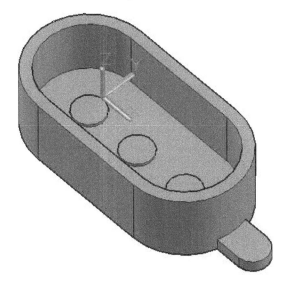

30. On the ribbon, click **Home** tab > **Coordinates** panel > **View** drop-down > **Face**.

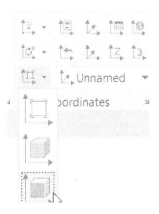

31. Click on the top face of the newly created cylinder.

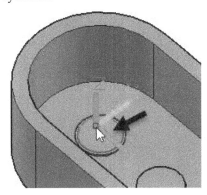

32. Press ENTER to accept the selection. The UCS is placed on the selected face.
33. On the ribbon, click **Home** tab > **Modeling** panel > **Primitives** drop-down > **Cylinder**.
34. Place the pointer on the circular edge of the cylinder; the centerpoint of the circular edge is displayed.
35. Select the centerpoint of the circular edge.

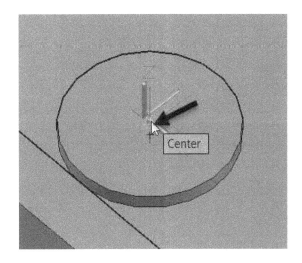

119

36. Type 16 and press ENTER.
37. Move the pointer downward and click outside the model.

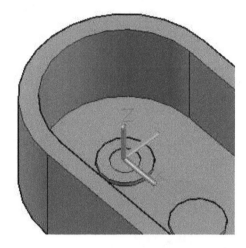

38. Select the newly created cylinder.
39. On the ribbon, click **Home** tab > **Modify** panel > **Array** drop-down > **Rectangular Array**.
40. On the **Array Creation** tab of the ribbon, type **3,** and **1** in the **Columns** and **Rows** boxes, respectively.
41. Type **110** in the **Between** box on the **Columns** panel (enter a negative value if the array is displayed in the reverse direction).
42. Click the **Close Array** icon.

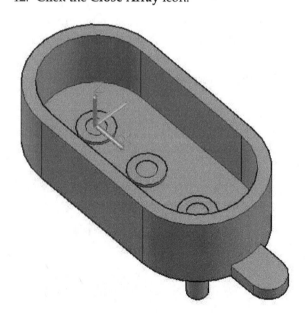

43. On the ribbon, click **Home** tab > **Modeling** panel > **Primitives** drop-down > **Cylinder**.
44. Place the pointer on the circular edge on the right side; the centerpoint of the circular edge is displayed.
45. Select the centerpoint of the circular edge.
46. Type 10 and press ENTER.
47. Move the pointer downward and click outside the model.

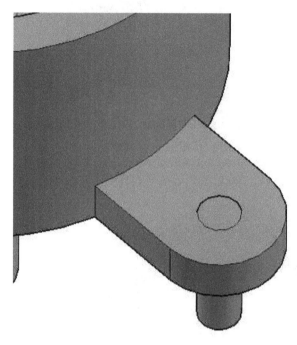

48. On the ribbon, click **Home** tab > **Modeling** panel > **Primitives** drop-down > **Cylinder**.
49. Press and hold the SHIFT key.
50. Right-click and select the **Mid Between 2 Points** option.
51. Select the vertices of the model, as shown.

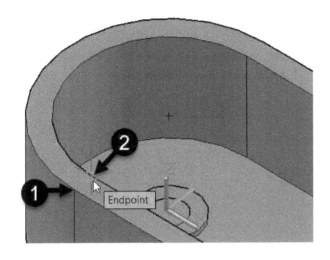

52. Type 5 and press ENTER.
53. Type -50.8 and press ENTER.

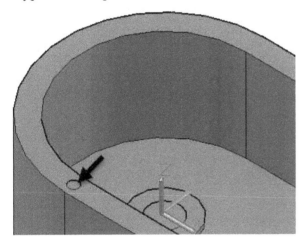

54. Change the **Visual Style** to **2D Wireframe**.
55. On the ribbon, click **Home** tab > **Selection** panel > **Filter** drop-down > **Edge**.

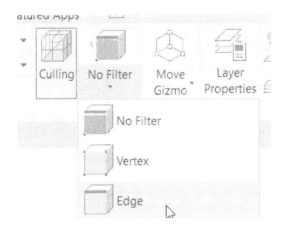

56. On the ribbon, click **Home** tab > **Solid Editing** panel > **Edge** drop-down > **Extract Edges**.

57. Select the edges of the model, as shown.

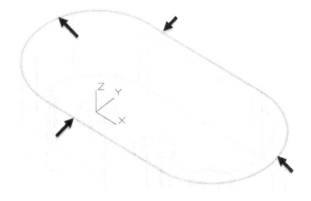

58. Press ENTER.
59. On the ribbon, click the **Home** tab > **Modify** panel > **Offset**.
60. Type 10 and press ENTER.
61. Select the extracted horizontal edge.
62. Move the pointer inside the model and click.
63. Likewise, offset the other edges, as shown.

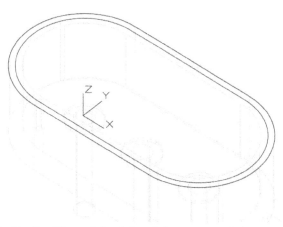

64. On the **Home** tab of the ribbon, expand the **Modify** panel and click the **Edit Polyline** icon.

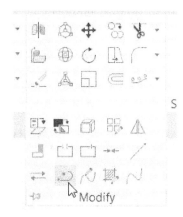

65. Select the **Multiple** option from the command line.
66. Select all the offset entities and press ENTER.

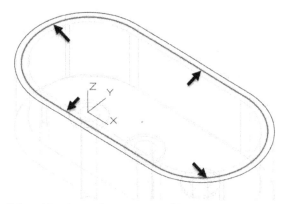

67. Select **Yes** from the command line.
68. Select the **Join** option from the command line.
69. Press ENTER to specify 0 as the fuzz distance.
70. Press ESC.

71. On the ribbon, click **Home** tab > **Selection** panel > **Filter** drop-down > **No Filter**.
72. On the ribbon, click **Home** tab > **Modify** panel > **Array** drop-down > **Path Array**.

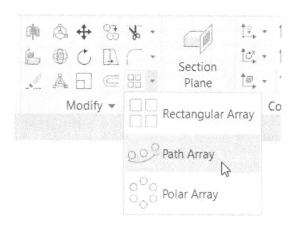

73. Select the cylinder and press ENTER.

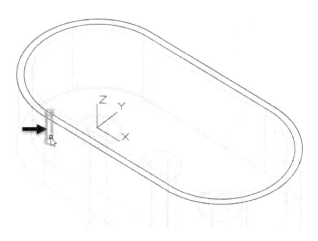

74. Select the polyline created using the **Edit Polyline** command.

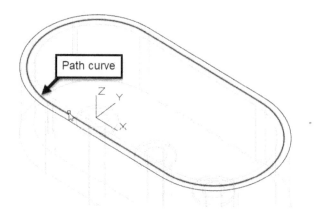

75. Click the **Base Point** icon on the **Properties** panel.

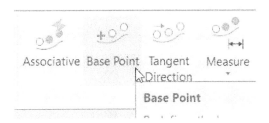

76. Select the centerpoint of the cylinder to be arrayed.

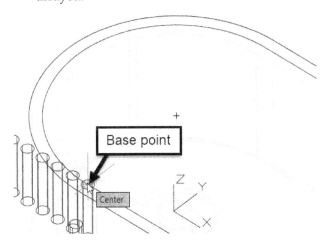

77. Click the **Item Count** icon on the **Items** panel.

78. Type **12** in the **Items** box on the **Items** panel.

79. Click **Measure Method** drop-down > **Divide** on the **Properties** panel.

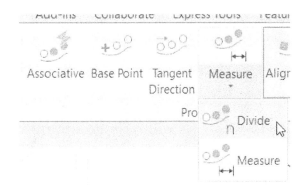

80. Click the **Close Array** icon.
81. Select the polyline and the extracted edges.
82. Press DELETE.

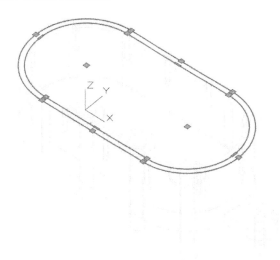

83. On the ribbon, click **Home** tab > **Solid Editing** panel > **Solid, Subtract**.

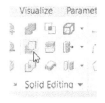

84. Select the main body and press ENTER.

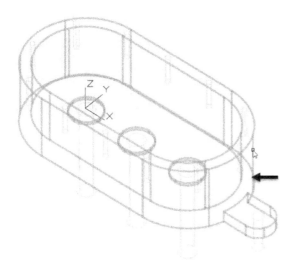

85. Select all the cylinders and press ENTER.

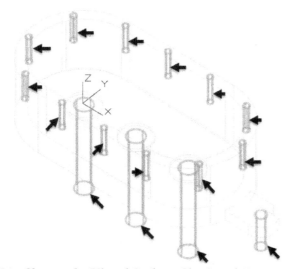

86. Change the **Visual Style** to **Shades of Gray**.

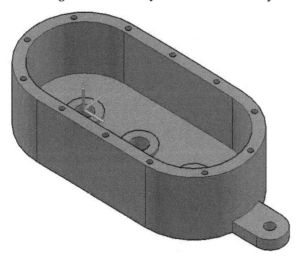

Creating Edge Fillets

1. On the ribbon, click **Solid** tab > **Solid Editing** panel > **Fillet Edge**.
2. Click on the vertical edges of the geometry, as shown.

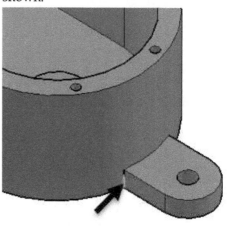

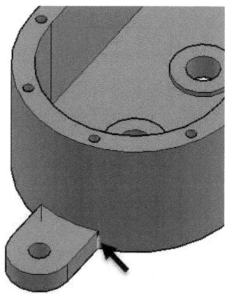

3. Select the **Radius** option from the command line.
4. Type 20 and press ENTER thrice.

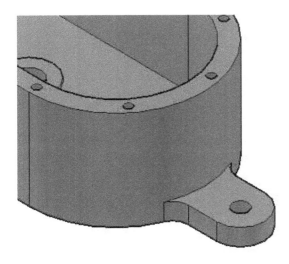

5. On the ribbon, click **Solid** tab > **Solid Editing** panel > **Fillet Edge**.
6. Click on the circular edges of the geometry, as shown.

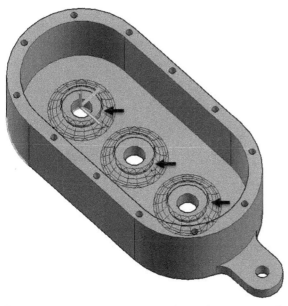

7. Select the **Radius** option from the command line.
8. Type 5 and press ENTER thrice.

Moving Faces

1. On the ribbon, click the **Home** tab > **Solid Editing** panel > **Solid Editing** drop-down > **Move Faces**.

2. Select the cylindrical faces of the model, as shown.

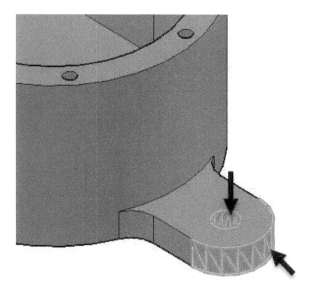

3. Press ENTER.
4. Select the endpoint of the circular edge.

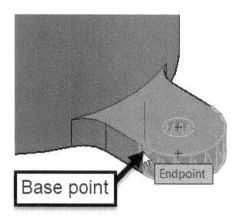

5. Move the pointer backward.

6. Type 20 and press ENTER.

7. Select the **Move** option from the command line.
8. Select the top face of the model.

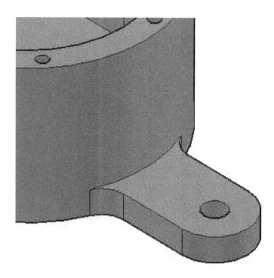

9. Press ENTER.
10. Select the endpoint of the vertical edge, as shown.

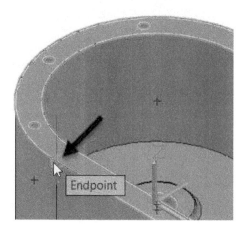

11. Move the pointer downward.

12. Type 40 and press ENTER.

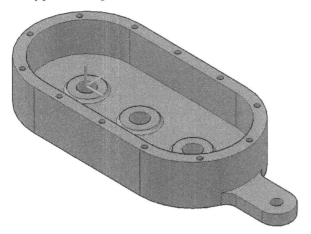

13. Click the **Orbit** tool on the Navigation Bar located at the right of the graphics window.

14. Press and hold the left mouse button and drag the pointer downward; the top portion is displayed.
15. Right-click and select **Exit**.
16. Select the **Move** option from the command line.
17. Select the bottom faces of the holes.
18. Press ENTER.

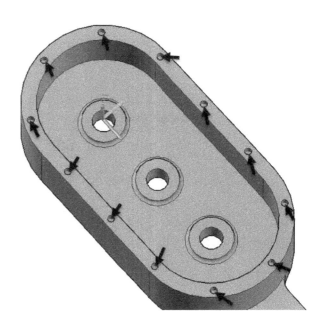

19. Select the endpoint of the vertical edge, as shown.

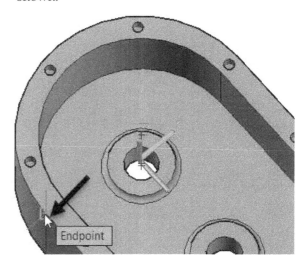

20. Move the pointer downward.
21. Type 40 and press ENTER.
22. Save and close the file.

Exercises
Exercise 1

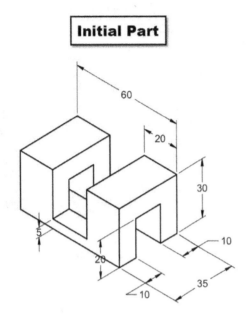

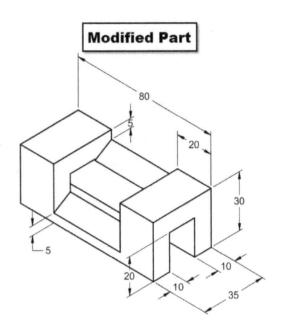

Exercise 2

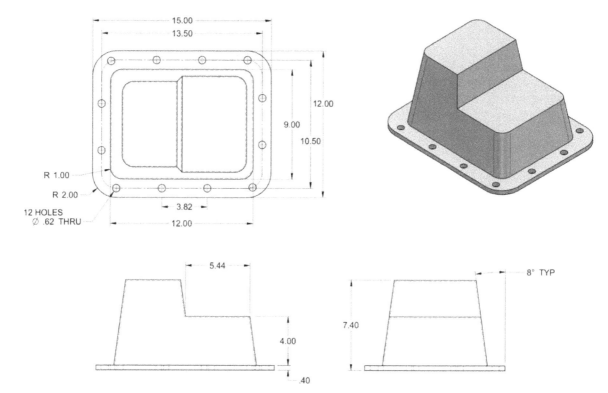

15.00
13.50
12.00
9.00
10.50
R 1.00
R 2.00
12 HOLES
Ø .62 THRU
3.82
12.00

5.44
4.00
.40

8° TYP
7.40

SHEET THICKNESS = 0.079 in

Chapter 9: Assemblies

Tutorial 1

In this example, you create the assembly shown below.

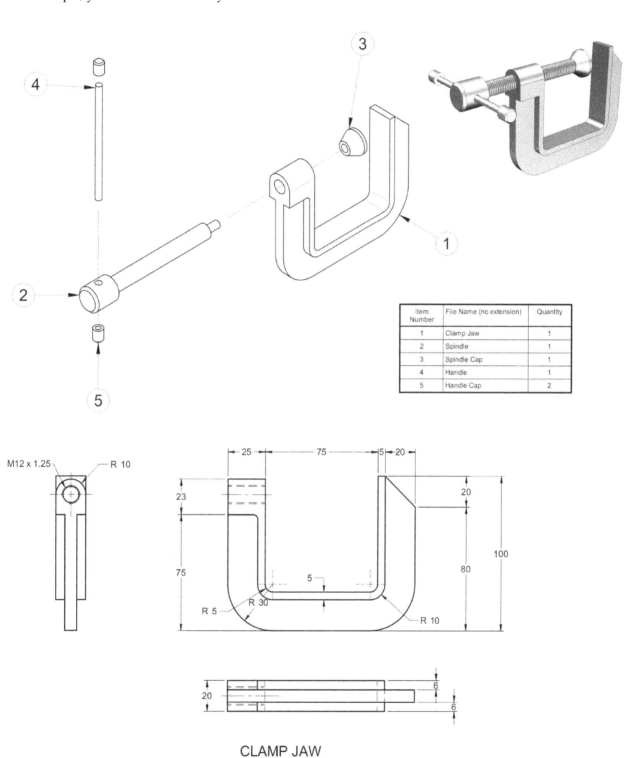

Item Number	File Name (no extension)	Quantity
1	Clamp Jaw	1
2	Spindle	1
3	Spindle Cap	1
4	Handle	1
5	Handle Cap	2

CLAMP JAW

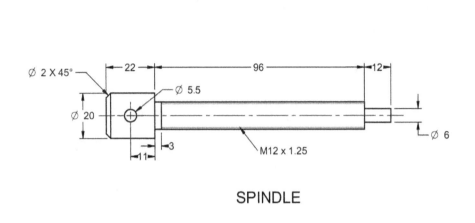

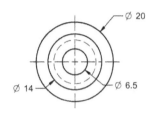

SPINDLE

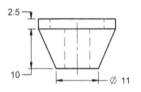

SPINDLE CAP

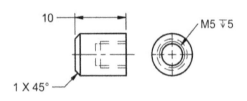

HANDLE CAP

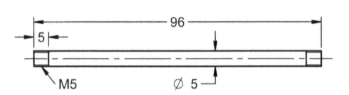

HANDLE

Creating a New File

1. Click the **New** icon on the Quick Access Toolbar.

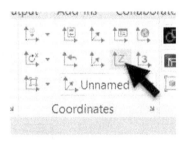

2. Select the **acadiso3D** template from the **Select template** dialog.
3. Click the **Open** button.
4. Deactivate the **GRIDMODE** icon on the status bar.

Creating the Clamp Jaw

1. On the **Home** tab of the ribbon, click **Coordinates > Z-Axis Vector**.

2. Press ENTER to specify 0,0,0 as the origin of the UCS, as shown.
3. Activate the Orthomode icon on the status bar.
4. Move the pointer toward the left and click. The Z-axis of the UCS is defined.

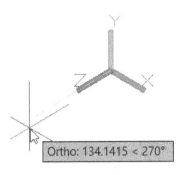

Ortho: 134.1415 < 270°

5. On the ribbon, click **Home** tab > **Modeling** panel > **Primitives** drop-down > **Cylinder**.
6. Type 0,88 and press ENTER to define the centerpoint if the cylinder.
7. Type 10 and press ENTER to define the radius of the cylinder.
8. Type -25 and press ENTER.

9. On the ribbon, click **Home** tab > **Modeling** panel > **Primitives** drop-down > **Box**.
10. Select the left quadrant point of the cylinder, as shown.

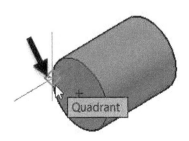

Quadrant

11. Select the **Length** option from the command line.
12. Move the pointer toward the right and select the right quadrant point.

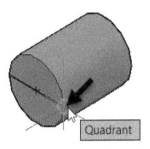

Quadrant

13. Move the pointer downward.
14. Type 13 and press ENTER
15. Move the pointer toward the right and select the quadrant point on the back face of the cylinder.

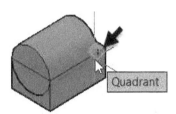

Quadrant

16. On the ribbon, click **Home** tab > **Solid Editing** panel > **Solid, Union**.
17. Select the cylinder and box, and then press ENTER.

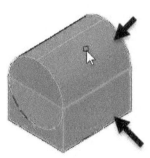

Creating the Swept Solid

1. On the **Home** tab of the ribbon, click **Coordinates > Z-Axis Vector**.

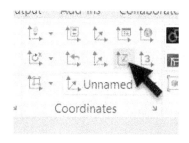

2. Orbit the model and select the midpoint of the bottom-front edge.

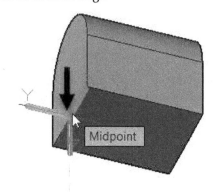

3. Move the pointer downward and click to specify the Z-axis.

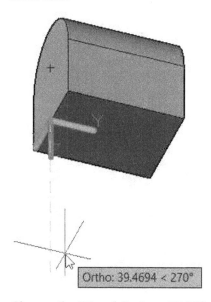

4. Change the **Visual Style** to **2D Wireframe**.
5. On the ribbon, click **Home** tab > **Selection** panel > **Filter** drop-down > **Edge**.
6. On the ribbon, click **Home** tab > **Solid Editing** panel > **Edge** drop-down > **Extract Edges**.
7. Select the edges of the bottom face of the model, as shown.

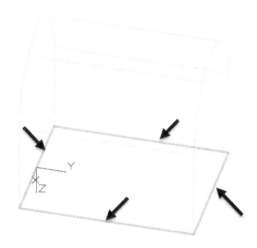

8. Press ENTER.
9. On the ribbon, click **Home** tab > **Selection** panel > **Filter** drop-down > **No Filter**.
10. On the ribbon, click the **Home** tab > **Modify** panel > **Offset**.
11. Type 5 in the command line and press ENTER. The offset distance is defined.
12. Select the extracted edge, as shown.
13. Move the pointer inside the model and click.

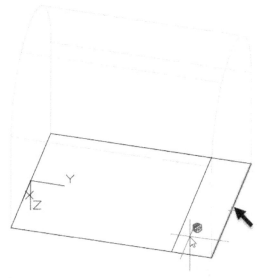

14. Press ENTER twice.
15. Type 6 and press ENTER.
16. Select the extracted edge, as shown.

134

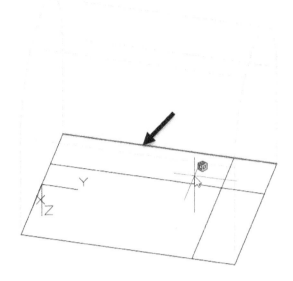

17. Move the pointer inside the model and click.
18. Select the extracted edge, as shown.
19. Move the pointer inside the model and click.

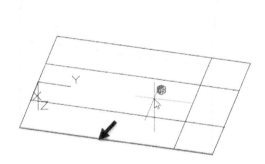

20. On the ribbon, click **Home** tab > **Modify** panel > **Trim/Extend** drop-down > **Trim**.
21. Select the portions of the horizontal line extending outside the model.

22. Press ESC.

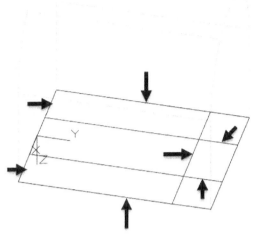

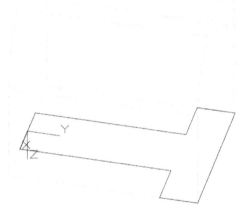

23. On the **Home** tab of the ribbon, click **Coordinates > Z-Axis Vector**.

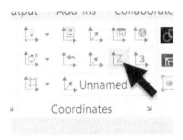

24. Orbit the model and select the midpoint of the bottom back edge.

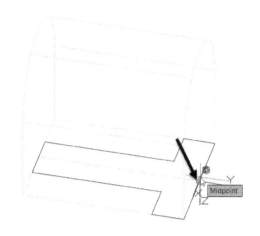

25. Select the corner point of the model, as shown.

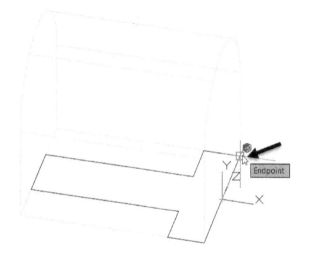

26. Change the View orientation to SE Isometric.
27. On the ribbon, click **Home** tab > **Draw** panel > **Polyline**.
28. Type 0,0 in the command line and press ENTER.
29. Turn ON the Dynamic Input and ORTHOMODE icons on the status bar.
30. Move the pointer downward.
31. Type 50 and press ENTER.

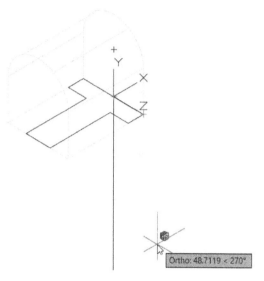

32. Move the pointer toward the right.

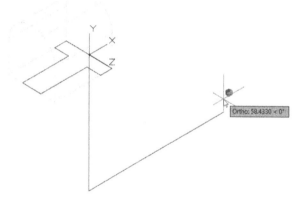

33. Type 75 and press ENTER.
34. Move the pointer upward.

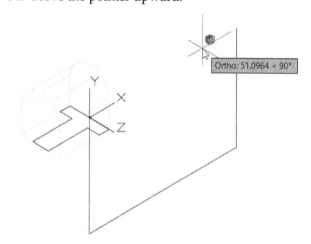

35. Type 75 and press ENTER.
36. Press ESC.

37. On the ribbon, click **Home** tab > **Modify** panel >
    Fillet drop-down > **Fillet**.

Modify ▾

38. Select the **Radius** option from the command
    line.
39. Type **5** in the command line and press ENTER.
40. Select Vertical and horizontal lines meeting at
    the corner, as shown.

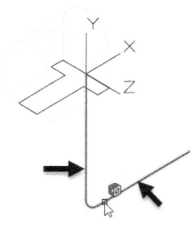

41. Press ENTER to activate the Fillet command.
42. Select Vertical and horizontal lines meeting at
    the corner, as shown.

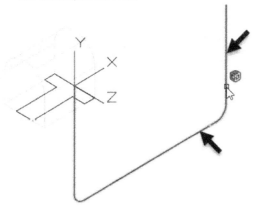

43. On the **Home** tab of the ribbon, expand the
    Draw panel and click the **Region** icon.
44. Select all the entities, as shown.

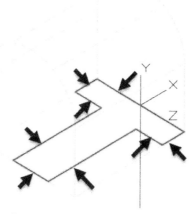

45. Press ENTER to convert all the 2D elements into
    a region.
46. On the ribbon, click the **Home** tab > **Modeling**
    panel > **Solids** drop-down > **Sweep**.
47. Select the region and press ENTER to define the
    object to sweep.

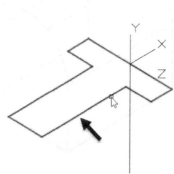

48. Select the polyline and press ENTER to define
    the path.

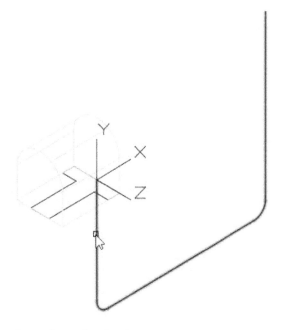

Creating the Hole

1. Change the **Visual Style** to **Shades of gray**.
2. On the ribbon, click **Home** tab > **Coordinates** panel > **Z-Axis Vector**.
3. Select the centerpoint of the circular edge, as shown.

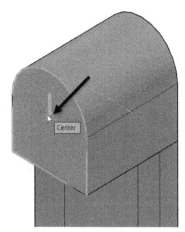

4. Move the pointer toward the left and click to define the Z-axis.

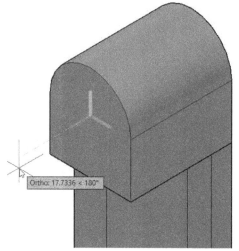

5. On the ribbon, click **Home** tab > **Modeling** panel > **Primitives** drop-down > **Cylinder**.
6. Select the centerpoint of the circular edge, as shown.

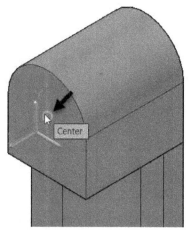

7. Type 6 and press ENTER to define the radius of the cylinder.
8. Move the pointer toward the right and click.

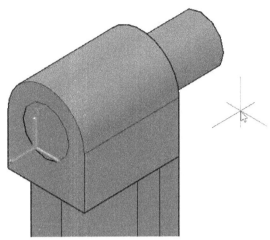

9. On the ribbon, click **Home** tab > **Solid Editing** panel > **Solid, Subtract**.
10. Select the body, as shown.

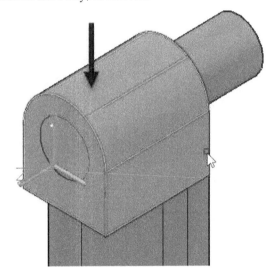

11. Press ENTER.
12. Select the cylinder and press ENTER.

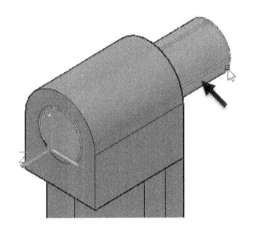

Creating the Chamfers and Fillets

1. On the ribbon, click the **Solid** tab > **Solid Editing** panel > **Fillet Edge** drop-down > **Chamfer Edge**.
2. Select the horizontal edge, as shown.

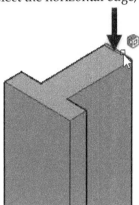

3. Select the **Distance** option from the command line.
4. Type **20** in the command line and press ENTER. Distance 1 is defined.
5. Type **20** in the command line and press ENTER. Distance 2 is defined.
6. Press ENTER twice to create the chamfer.

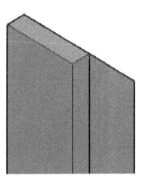

7. On the ribbon, click **Home** tab > **Solid Editing** panel > **Solid, Union**.
8. Select all the bodies and press ENTER.
9. On the ribbon, click **Solid** tab > **Solid Editing** panel > **Fillet Edge**.
10. Click the bottom front corner of the ViewCube; the view orientation is changed.

11. Select the inner horizontal edge of the model, as shown.

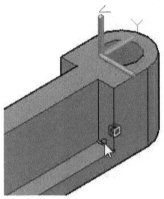

12. Select the top-left corner of the ViewCube. The orientation of the model is changed.

13. Select the visible inner horizontal edge.

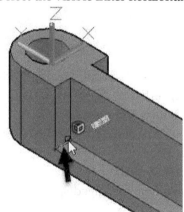

14. Select the **Radius** option from the command line.
15. Type 2 and press ENTER thrice.

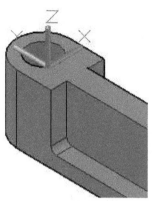

16. Change the View orientation to SE Isometric.

Creating the Spindle

1. On the ribbon, click **Home** tab > **Modeling** panel > **Primitives** drop-down > **Cylinder**.
2. Select the centerpoint of the circular edge, as shown.

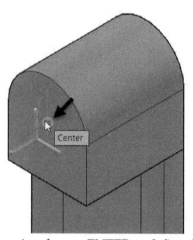

3. Type 6 and press ENTER to define the radius of the cylinder.
4. Type -96 and press ENTER.

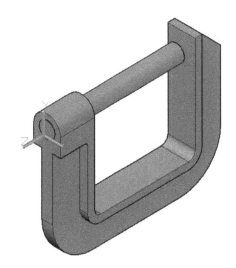

5. Select the newly created cylinder.
6. Select the Y-axis of the move gizmo.

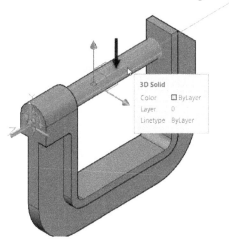

7. Move along the selected axis toward left.

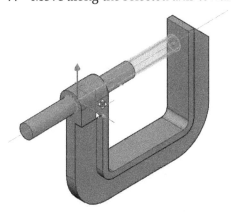

8. Type 16 and press ENTER.
9. On the ribbon, click **Home** tab > **Modeling** panel > **Primitives** drop-down > **Cylinder**.
10. Select the centerpoint of the front face of the cylinder.

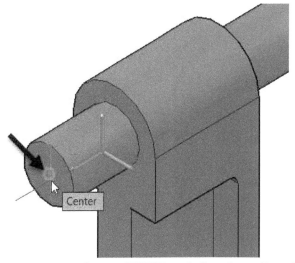

11. Type 10 and press ENTER to define the radius of the cylinder.
12. Type 22 and press ENTER to define the height of the cylinder.

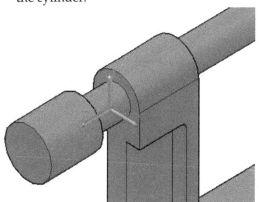

13. Change the View orientation to **NW Isometric**.

14. On the ribbon, click **Home** tab > **Modeling** panel > **Primitives** drop-down > **Cylinder**.
15. Select the centerpoint of the end face of the cylinder.

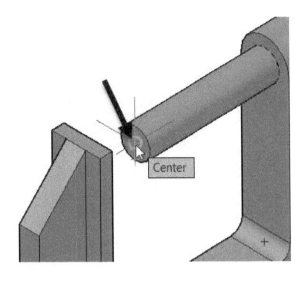

16. Type 3 and press ENTER to define the radius of the cylinder.
17. Type -12 and press ENTER to define the height of the cylinder.

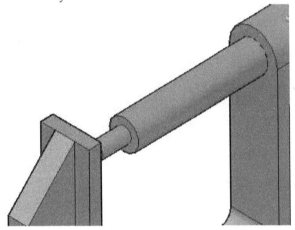

18. Change the View orientation to **SE Isometric**.
19. On the ribbon, click the **Solid** tab > **Solid Editing** panel > **Fillet Edge** drop-down > **Chamfer Edge**.
20. Select the circular edge of the cylinder.

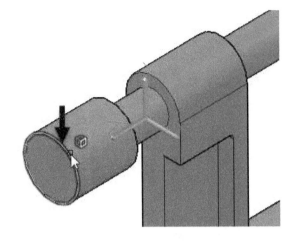

21. Select the **Distance** option from the command line.
22. Type **2** in the command line and press ENTER. Distance 1 is defined.
23. Type **2** in the command line and press ENTER. Distance 2 is defined.
24. Press ENTER twice to create the chamfer.

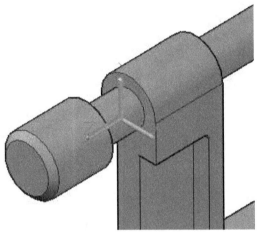

25. On the **Home** tab of the ribbon, click **Coordinates > Z-Axis Vector**.

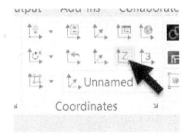

26. Select the quadrant point of the circular edge, as shown.

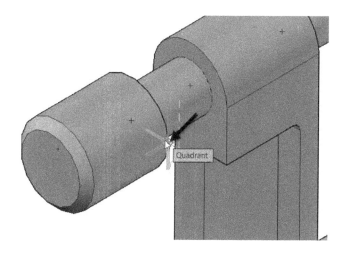

27. Move the pointer backward and click. The Z-axis of the UCS is defined.

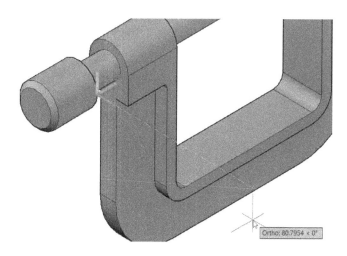

28. On the ribbon, click **Home** tab > **Modeling** panel > **Primitives** drop-down > **Cylinder**.
29. Type -11,0 and press ENTER to define the centerpoint of the cylinder.
30. Select the **Diameter** option from the command line.
31. Type 5.5 in the command line and press ENTER.
32. Move the pointer toward the left and click outside the model.

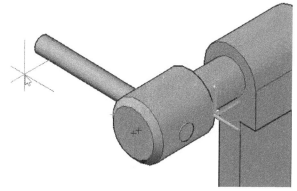

33. On the ribbon, click **Home** tab > **Solid Editing** panel > **Solid, Subtract**.
34. Select the large cylinder and press ENTER.
35. Select the small cylinder and press ENTER.

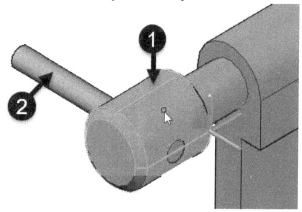

36. On the ribbon, click **Home** tab > **Solid Editing** panel > **Solid, Union**.
37. Select the three cylinders and press ENTER.

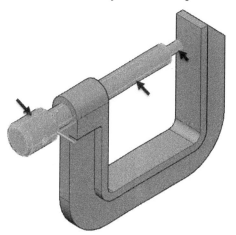

Creating the Spindle Cap

1. Change the View orientation to NW Isometric.
2. On the **Home** tab of the ribbon, click **Coordinates > Z-Axis Vector**.

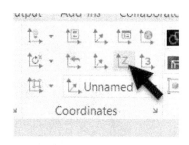

3. Select the center point of the circular edge, as shown.

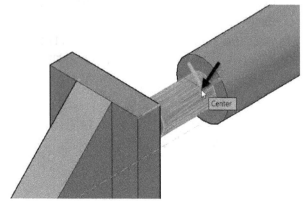

4. Move the pointer toward the left and click. The Z-axis of the UCS is defined.

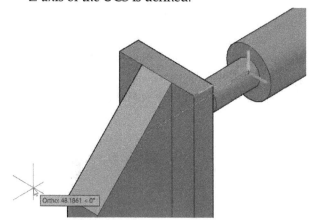

5. On the ribbon, click **Home** tab > **Modeling** panel > **Primitives** drop-down > **Cylinder**.
6. Select the centerpoint of the circular edge, as shown.

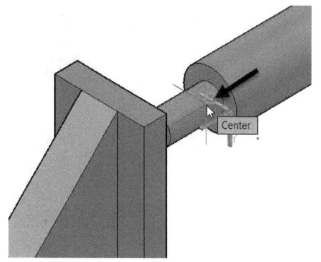

7. Select the **Diameter** option from the command line.
8. Type 11 and press ENTER to define the radius of the cylinder.
9. Type 10 and press ENTER.

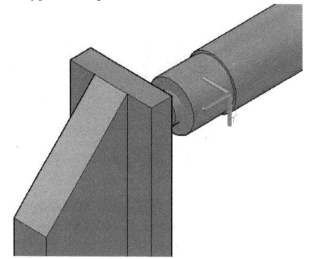

10. On the **Solid** tab of the ribbon, click **Solid Editing** panel > **Taper Faces**.

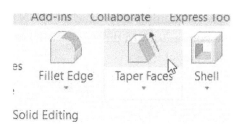

11. Select the cylindrical face, as shown.

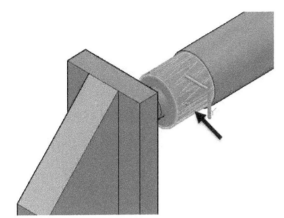

12. Press ENTER.
13. Select the centerpoint of the circular edge, as shown.

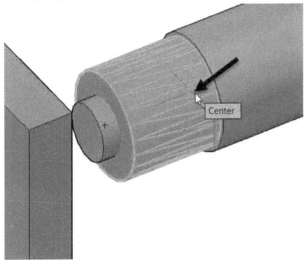

14. Move the pointer in the Z direction of the UCS, and then click to specify the axis.

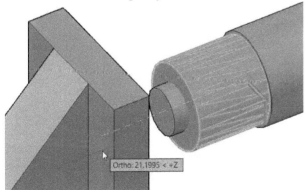

15. Type -24.228 and press ENTER to specify the taper angle.

16. Select the **Extrude** option from the command line.
17. Select the end face of the spindle cap.

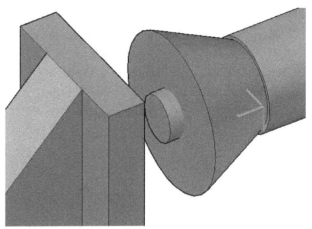

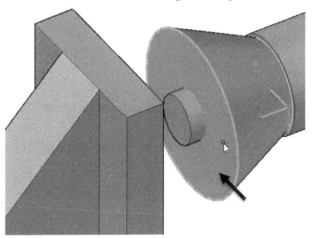

18. Press ENTER
19. Type 2.5 and press ENTER.
20. Press ENTER.

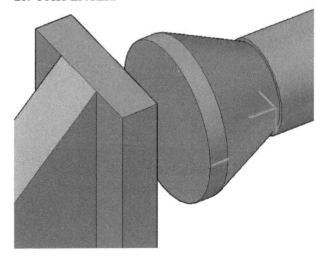

21. Press ESC.
22. Deactivate the **Dynamic Input** icon on the status bar.
23. On the ribbon, click **Home** tab > **Modeling** panel > **Primitives** drop-down > **Cylinder**.
24. Select the centerpoint of the circular edge, as shown.

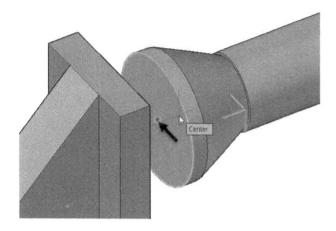

25. Select the **Diameter** option from the command line.
26. Type 6.5 in the command line and press ENTER.
27. Type -12.5 in the command line and press ENTER.

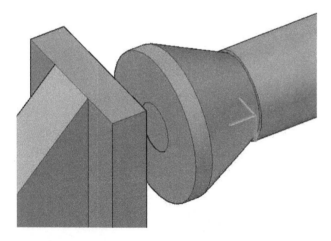

28. On the ribbon, click **Home** tab > **Modeling** panel > **Primitives** drop-down > **Cylinder**.
29. Select the centerpoint of the circular edge, as shown.

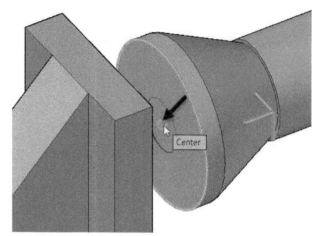

30. Type 7 in the command line and press ENTER.
31. Type -2.5 in the command line and press ENTER.

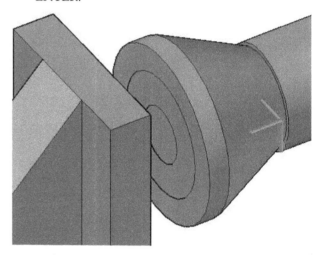

32. On the ribbon, click **Home** tab > **Solid Editing** panel > **Solid, Subtract**.
33. Select the tapered cylinder and press ENTER.

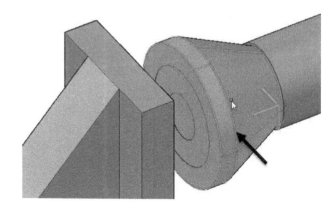

34. Select the two cylinders inside the tapered cylinder, as shown.

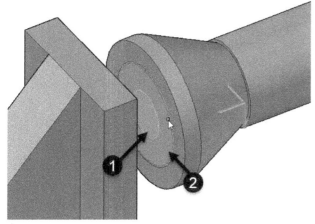

35. Press ENTER.

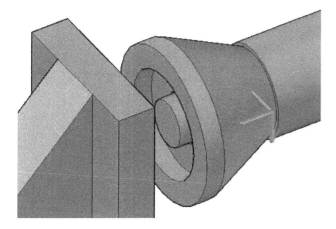

Creating the Handle

1. Change the View orientation to **SE Isometric**.
2. On the ribbon, click **Home** tab > **Coordinates** panel > **UCS, Previous**. The UCS is moved to its previous location.

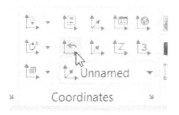

3. On the ribbon, click **Home** tab > **Modeling** panel > **Primitives** drop-down > **Cylinder**.
4. Type -11,0 and press ENTER to define the centerpoint of the cylinder.

5. Select the **Diameter** option from the command line.
6. Type 5 in the command line and press ENTER.
7. Type 96 and press ENTER.

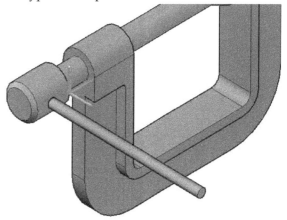

8. Select the newly created cylinder.
9. Click on the X-axis of the move gizmo and move toward left.

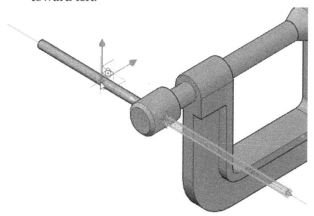

10. Type 58 and press ENTER.

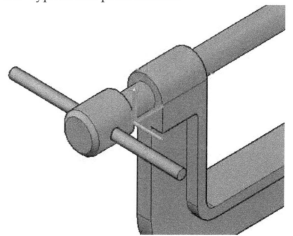

Creating the Handle Cap

1. On the ribbon, click **Home** tab > **Modeling** panel > **Primitives** drop-down > **Cylinder**.
2. Select the centerpoint of the circular edge, as shown.

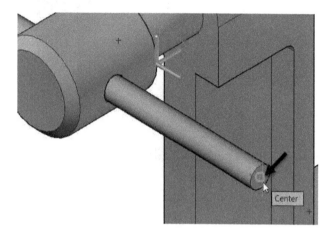

3. Select the **Diameter** option from the command line.
4. Type 8 in the command line and press ENTER.
5. Type 10 and press ENTER.

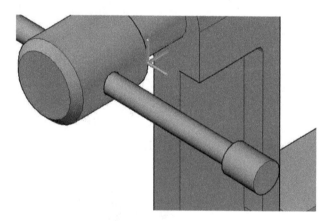

6. On the ribbon, click **Home** tab > **Modeling** panel > **Primitives** drop-down > **Cylinder**.
7. Select the centerpoint of the circular edge, as shown.

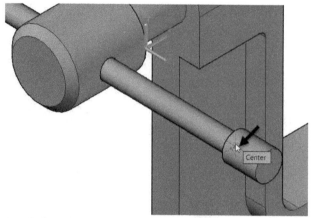

8. Select the **Diameter** option from the command line.
9. Type 5 in the command line and press ENTER.
10. Type 5 and press ENTER.
11. Change the **Visual Style** to **2D Wireframe**.
12. On the ribbon, click **Home** tab > **Solid Editing** panel > **Solid, Subtract**.
13. Select the large cylinder, as shown.

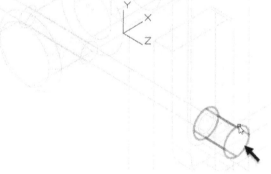

14. Press ENTER.
15. Select the small cylinder, as shown.

16. Press ENTER.
17. Change the **Visual Style** to **Shades of Gray**.
18. Select the handle cap.
19. Click on the X-axis of the Move gizmo.

20. Move along the selected axis toward left.
21. Type 5 and press ENTER.

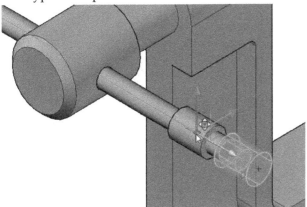

22. On the ribbon, click the **Solid** tab > **Solid Editing** panel > **Fillet Edge** drop-down > **Chamfer Edge**.
23. Select the circular edge of the cylinder.

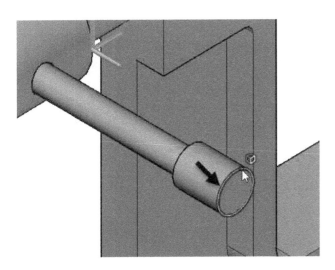

24. Select the **Distance** option from the command line.
25. Type **1** in the command line and press ENTER. Distance 1 is defined.
26. Type **1** in the command line and press ENTER. Distance 2 is defined.
27. Press ENTER twice to create the chamfer.

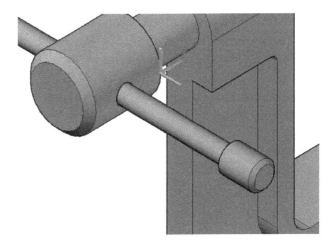

28. On the ribbon, click **Home** tab > **Coordinates** panel > **UCS, World**.

29. Select the handle cap.

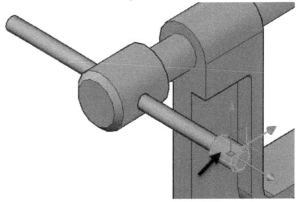

30. On the ribbon, click **Home** tab > **Modify** panel > **3D Mirror**.

31. Select the **YZ** option from the command line.
    The orientation of the mirror plane is defined
32. Press ENTER to define 0,0,0 as the location of
    the mirror plane.
33. Select **No** from the command line to keep the
    original object.

34. Save and close the drawing file.

Exercises
Exercise 1

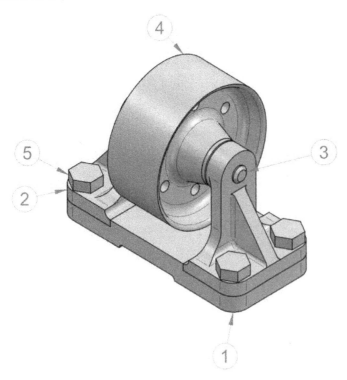

Item Number	File Name (no extension)	Quantity
1	Base	1
2	Bracket	2
3	Spindle	1
4	Roller-Bush assembly	1
5	Bolt	4

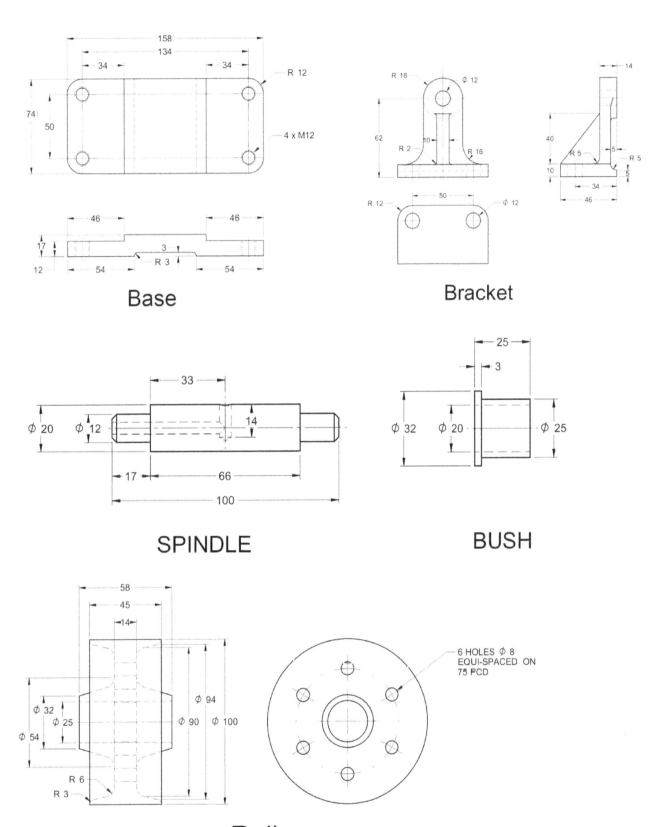

Base

Bracket

SPINDLE

BUSH

Roller

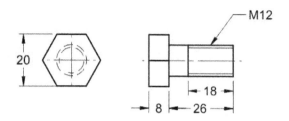

Bolt

Chapter 10: Drawings

Tutorial 1

In this example, you create the 2D drawing of the part shown below.

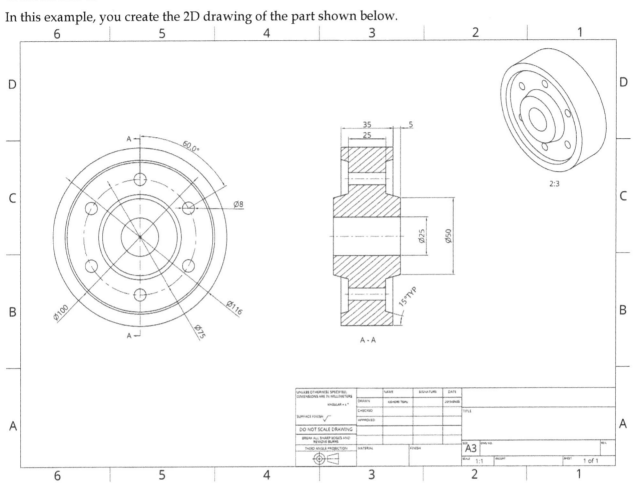

Creating the Drawing from a 3D Model

1. Download the **Chapter 10** part file from the companion website. Next, extract the zip file.
2. Open the Ch10_Tutorial1 file.

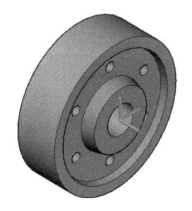

3. Click the right mouse button on the **Layout 1** tab.
4. Select the **Drafting Standard Setup** option.

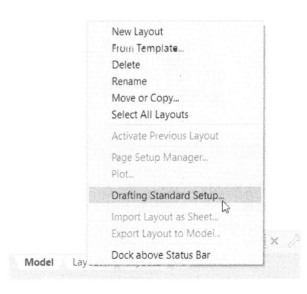

5. On the **Drafting Standard** dialog, select the **Third angle** option from the **Projection type** section.
6. Click **OK**.

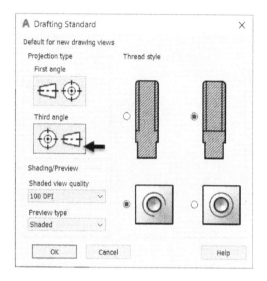

7. Click the **Layout 1** tab, and then select the **Page Setup Manager** option.
8. Select the **Layout 1** and click the **Modify** button.
9. On the **Page Setup** dialog, select **Paper size > ISO A3 (420.00 x 297.00 MM)**.
10. Click **OK**.
11. Close the **Page Setup Manager** dialog.
12. Select the viewport on the layout and press DELETE.

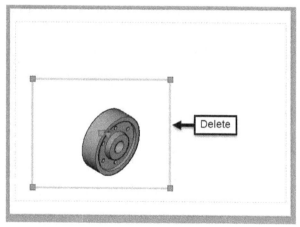

13. Click the **Model** tab.
14. On the ribbon, click **Home** tab > **View** panel > **Base** drop-down > **From Model Space**.

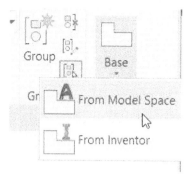

15. Press ENTER to select the entire model.
16. Press ENTER to select **Layout 1** as the current layout.
17. On the **Drawing View Creation** tab of the ribbon, select the **Right** option from the **Orientation** panel.

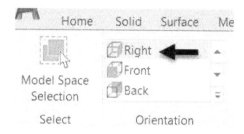

18. On the **Appearance** panel, set the **Scale** to **1:1**.

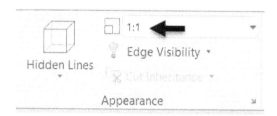

19. Select the **Visible Lines** option from the drop-down available on the **Appearance** panel.

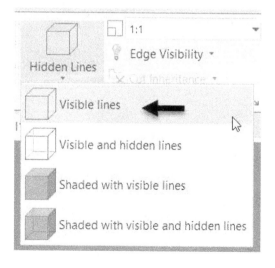

20. Click on the left side of the layout, as shown.

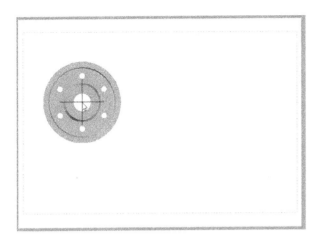

21. Click **OK** on the **Create** panel.

22. Move the pointer toward the top left corner and click to position the Isometric view, as shown.

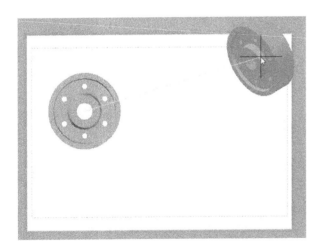

23. Press ENTER.
24. Select the Isometric view.
25. Click on the square grip of the select view.
26. Move the pointer inside the layout boundary and click.

27. Select the isometric view.
28. On the ribbon, click **Edit > Edit View**.

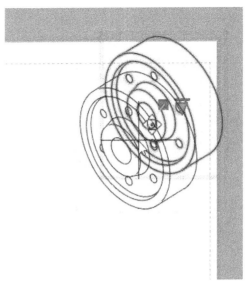

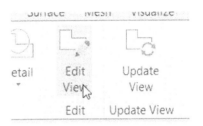

29. Type **2:3** in the **Scale** box on the **Appearance** panel.
30. Click the **OK** button.

31. Move the Isometric to the top right corner.

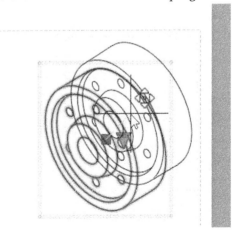

Creating the Section view

1. On the **Layout** tab of the ribbon, click **Create New** panel > **Section View** drop-down > **Full**.

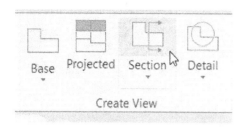

2. Select the Right view.
3. Place the pointer on the centerpoint of the selected view.

4. Move the pointer vertically upward. A dotted line appears from the center point of the selected view.
5. Click on the dotted line.

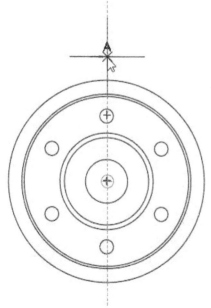

6. Move the pointer vertically downward.
7. Click below the selected view.

8. Press ENTER.

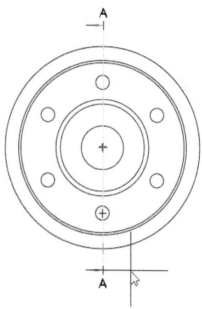

9. Move the pointer toward the right and click.

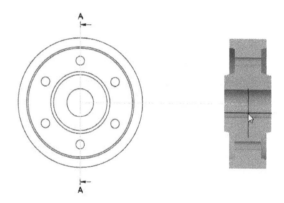

10. Click **OK** on the **Create** panel of the ribbon.

Adding Centerlines and Center Marks

1. On the ribbon, click **Home** tab > **Layers** panel > **Layer Properties**.
2. Click the **New Layer** icon on the Layer Properties Manager.
3. Type **Centerlines** in the **Name** box.
4. Change the layer **Color** to Black.
5. Set the Linetype to **CENTER**.
6. Make the layer current.

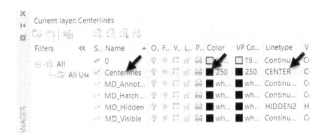

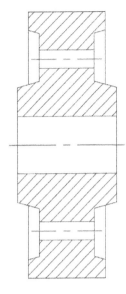

7. Close the Layer Properties Manager.

8. On the ribbon, click **Annotate** tab > **Centerlines** panel > **Centerline**.

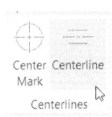

9. Select the horizontal edges of the large hole.

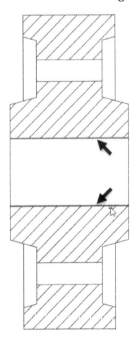

10. Likewise, create centrelines for the small holes.

11. On the ribbon, click **Home** tab > **Draw** panel > **Circle** drop-down > **3 Point**.

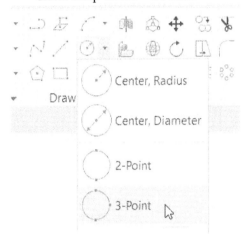

12. Select the centerpoints of the three circles, as shown.

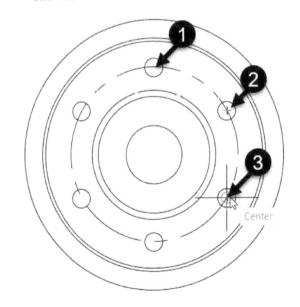

13. On the ribbon, click **Home** tab > **Draw** panel > **Line**.
14. Place the pointer on the lower quadrant point of the lower circle, as shown.
15. Move the pointer downward and click to specify the starting point of the line.

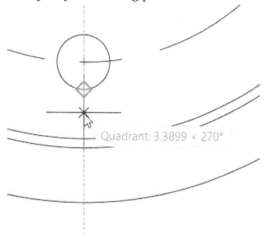

16. Move the pointer vertically upward and click to define the endpoint of the line.

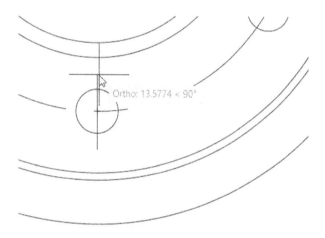

32. Press ESC.
33. On the ribbon, **Home** tab > **Modify** panel > **Array** drop-down > **Polar Array**.

34. Select the newly created line and press ENTER.

35. Select the centerpoint of the center circle; the centerpoint of the polar array is defined.

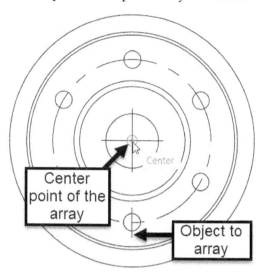

36. On the **Items** panel, type **6**, and **360** in the **Items** and **Fill** boxes, respectively.

37. Make sure that the **Rotate Items** icon is selected on the **Properties** panel.

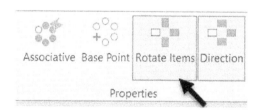

38. Click **Close Array**.

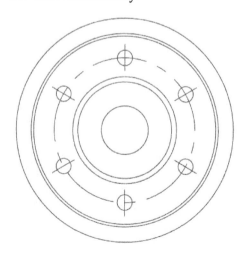

39. On the ribbon, click **Annotate** tab > **Centerlines** panel > **Center Mark**.

40. Select the center hole.

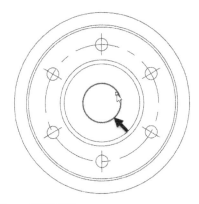

41. Press ENTER.

Adding Dimensions

Now, you add the dimensions to the drawing views.

1. Create the **Dimension** layer and make it as current.

2. Click **Dimension** drop-down > **Linear** on the **Dimension** panel of the **Annotate** ribbon tab.

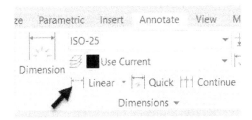

3. Select the two points of the section view, as shown.

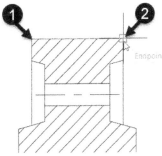

4. Move the pointer upwards and click to add dimension.

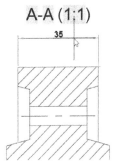

5. Click **Dimensions** panel > **Continuous** on the ribbon.

6. Select the corner point of the line of the section view, as shown.

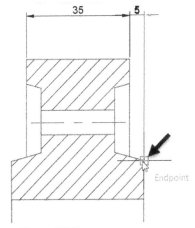

7. Press ESC.

8. Click **Dimensions** panel > **Dimension** on the **Annotate** tab of the ribbon.

9. Select the endpoints of the two parallel edges of the section view, as shown. Next, move the pointer upward and click to position the dimension.

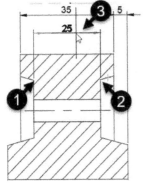

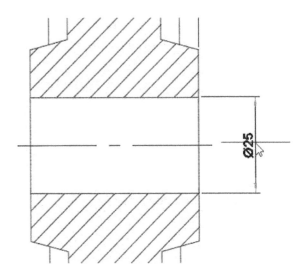

10. Click the **Dimension** icon on the **Dimensions** panel of the **Annotate** ribbon tab.

11. Select the endpoints of the two horizontal edges of the section view, as shown.

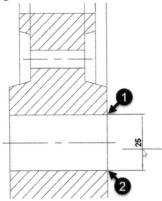

12. Select the **MText** option from the command line.

13. On the **Text Editor** tab of the ribbon, select Insert panel > **Symbol** drop-down > **Diameter**.

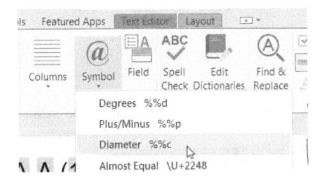

14. Click in the graphics window.

15. Move the pointer toward the right and click to position the dimension.

16. Press ESC.

17. Likewise, add another dimension to the section view, as shown.

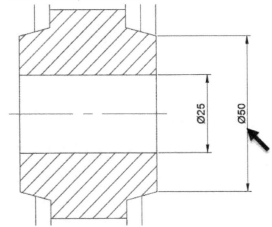

18. Click **Dimensions** panel > **Dimensions** dropdown > **Angular** on the **Annotate** tab of the ribbon.

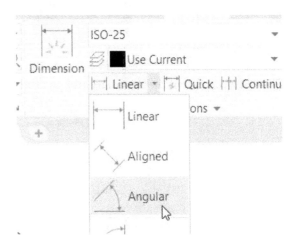

19. Select the two edges of the section view, as shown.

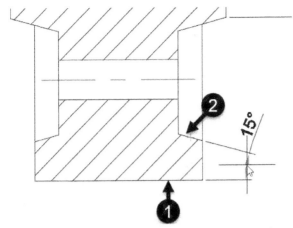

20. Select the **MText** option from the command line.
21. Expand the text editor by dragging the arrow displayed next to it.
22. Click next to the dimension value.
23. Type TYP and click in the graphics window.

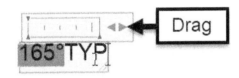

24. Move the pointer toward the right and click to position the angular dimension.

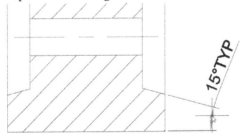

25. Click **Dimension** drop-down > **Diameter** on the **Dimensions** panel of the **Annotate** ribbon tab.

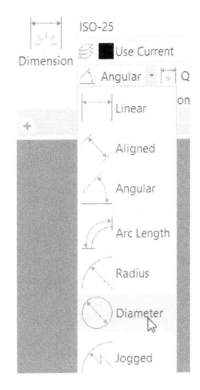

26. Select the outer circular edge, as shown.
27. Move the pointer and click to position the dimension at the location, as shown.

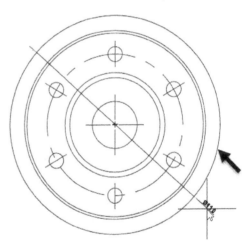

28. Likewise, create other diameter dimensions, as shown.

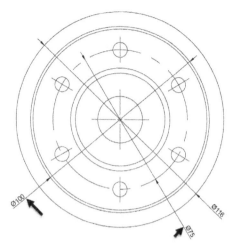

29. Click **Dimension** drop-down > **Diameter** on the **Dimensions** panel of the **Annotate** ribbon tab.
30. Select the small hole on the front view, as shown.
31. Move the pointer diagonally and click to place the diameter dimension.

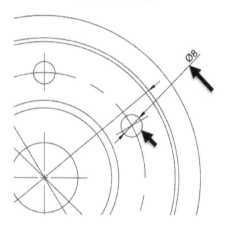

32. Click **Dimensions** panel > **Dimensions** drop-down > **Angular** on the **Annotate** tab of the ribbon.
33. Select the two centerlines of the circles, as shown.

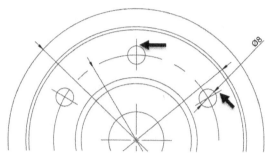

34. Move the pointer outward and click to position the angular dimension.

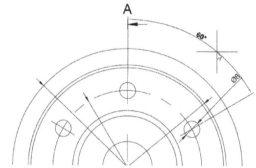

35. Save and close the drawing.

Exercise 2

Create orthographic views and an auxiliary view of the part model shown below. Add dimensions and annotations to the drawing.

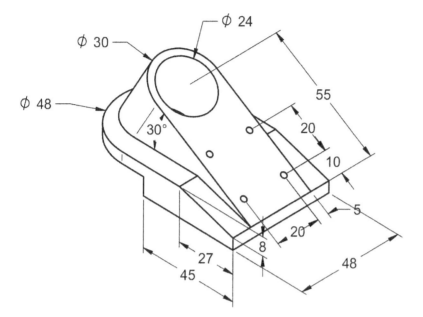

www.ingramcontent.com/pod-product-compliance
Lightning Source LLC
Chambersburg PA
CBHW080416060326

40689CB00019B/4269